THE PRACTICAL BUSHCRAFT SURVIVAL GUIDE

Essential Skills for Finding Food, Water, Shelter, and Fire in the Wilderness

Hunter Vale

TABLE OF CONTENTS

ISBN: 9781807652456

Published by TPSA Publishing

FOREWORD

I have been teaching hiking, camping, and backpacking for many years throughout many countries, but only within the last three years, a group of people in Columbia asked me to teach them survival tactics in the woods. That question opened my mind and eyes wide open; I know why they had asked me to teach them, but what if similar civil and or political unrest or war breaks out in our country, how many of us would be ready? More importantly how many of us would actually survive?

I believe every one of us needs to know at least the basics of wilderness survival. This way if you ever need it, you will have the knowledge. Along with the knowledge, let's make a survival pack ready just in case. In that pack, we can put a few necessary gears, and few "must have."s. Once you do these two things (first gather the basic knowledge of survival, second get the emergency pack ready), I believe you will sleep better at night.

When it comes to knowledge and gear, there is a big debate in the bushcraft community. Some believe that the true art of bushcraft is about going into the wilderness armed only with knowledge and a knife, everything else gets left at home.

On the other hand, there are those who believe bushcraft is about being prepared and bringing the necessary gear to ensure survival in the wilderness. Whatever your view is, I tried to cover both sides of that debate in this book. This way you are covered either way.

Now one thing to remember, I am not a magician, so just by words, I can't teach or show you how to light a fire without a lighter or a match, or how to get water from a tree trunk but I can tell you what you will need to get the task completed. I can also do something better, I can link a few videos from YouTube so as you read, you can also watch the videos and learn visually. Sounds better?

Let's get started....

What is Bushcraft?

A simple explanation for bushcraft is a set of skills to help you survive, live and thrive in the wilderness; often without amenities. Typically these skills are learned through knowledge and real life experience. However, bushcraft is about a lot more than simply surviving in the wilderness. It is also about sustainability and being comfortable while living in the natural environment; it is about living as a part of the natural world with a more indigenous approach.

Bushcraft is about learning more than just one thing. Rather it is a group of interrelated skills that allow you to adapt and overcome obstacles in order to survive in the wilderness. While traditional bushcraft is focused towards survival in the wilderness, the mindset of using what is available to you can also be applied to urban and suburban settings.

Using bushcraft skills in order to live off the land is a very satisfying experience. Connecting with nature allows you to reset both physically and mentally; nothing in modern life can compare with this. Anyone of any age can learn bushcraft skills. Just keep in mind that you are learning a skill that is done without any modern conveniences; it is just you and the wilderness.

After I had learned the skills of bushcraft I noticed five main benefits:

1. An increase in ability to adapt to challenges.
2. Improved self-sufficiency.
3. Improved confidence.
4. Increased survival skills.
5. Better prepared for unforeseen problems.

When I first started bushcraft learning, my first question was what skills would I be learning? This is something you need to know before knowing whether or not learning bushcraft is something that will interest you.

WHAT ARE BUSHCRAFT SKILLS?

When it comes to food foraging bushcraft focuses on the following four areas(I will get to the details little later in the book):

1. Learning about local plants.
2. Outdoor cooking methods.
3. How to avoid toxic plants.
4. Efficiently forage plants.

For trapping and hunting, bushcraft skills focus on the following eight areas:

1. Animal tracking.
2. Reading animal signs.
3. Building snares.
4. Fishing techniques.
5. Hiding human scent.
6. Knot tying.
7. Making cordage.
8. Cleaning and cooking food.

In the area of water gathering and purification, bushcraft teaches you the following five skills:

1. Finding water.
2. Making water filters.
3. Purifying water.
4. Building fire.
5. Making containers.

Within the area of building a fire, bushcraft skills also focuses on the following six areas:

1. Collecting wood.
2. Gathering tinder.
3. Batoning branches.
4. Building fire starting devices.
5. Building a fire pit.
6. Types of fire and their uses.

The last area of bushcraft skills involves shelter building and includes the following seven areas:

1. Felling trees.
2. Batoning branches.
3. Harvesting materials.
4. Thatching or weaving grass or other materials.
5. Knot tying.
6. Making cordage.
7. Natural insulation and waterproofing.

It is plain to see what bushcraft encompasses. However, bushcraft has recently been confused with wilderness survival in many ways and for good reasons too. Many see bushcraft as having a pack full of tools and heading out into the wilderness to survive.

Traditional bushcraft is not about that and when it comes to discussing what bushcraft is just remember modern day bushcraft has changed in many ways and now people are interested in learning how the bushcraft skills can make them live and survive in the woods when the time and the need arise.

What is Bushcraft Not About?

Perhaps the most important thing is that traditional bushcraft is about being responsible and leaving no trace. Bushcraft is not about camping in the wilderness and leaving evidence behind that you were there. You don't cut down live trees to build shelters or fires; that is destructive and unsustainable, something a true bushcraft practitioner wouldn't do.

The same can be said for shelters. True bushcraft wouldn't leave a shelter behind in the woods because it is an eyesore and also potentially dangerous.

Not only is it illegal to leave garbage around or even to bury it, but it is also someone wouldn't do when practicing bushcraft.

When lighting fires, bushcraft practicers do so in a way that doesn't leave charred ground with a ring of stones. This is disgraceful, potentially harmful to the environment and is certainly not something taught in bushcraft skills.

All of the above practices, when done incorrectly, would show disrespect and a lack of understanding of nature. This is not what bushcraft is about, which teaches an individual to live in harmony with nature.

On the other hand, a bushcraft practicer uses natural shelters that are built with dead material or in a way that won't impact the local environment. Shelters are always dismantled to return a campsite to a natural condition before leaving. For this reason, some choose to use a tarp.

Bushcraft also teaches safe fire practices and to make sure a fire is completely extinguished before leaving. If there are scorched areas, then it should be covered with fresh soil or debris, so no visual is left behind. You aren't a true bushcraft practicer unless you can see no remains of your campsite when you leave.

So now you know what bushcraft is about and what it can give you in benefits. The next thing is how you can be prepared to learn the art of bushcraft.

How to Get Prepared

If you decide to venture out into the wilderness, you are going to need to do some common sense preparation. You never want to simply head out into the wilderness without any planning.

However, there is always the chance something can go wrong; even on a daily hike. The key to surviving is your knowledge and ability to use bushcraft skills. Therefore, when it comes to getting prepared to learn bushcraft, you need to be aware of the survival basics.

SURVIVAL BASICS

When it comes to survival in any situation, there are four core components that you need to be aware of water, shelter, fire and food. Often in that order.

Although not in every case. For survival, you need to be clear about your priorities.

If you are looking at long-term survival after a natural disaster, then food is going to be more essential that shelter or sometimes fire, since food is something you can't live without. However, if you are lost in the wilderness, then exposure or attack from wild animals can make fire or shelter a bigger priority over food. Fire is also closely linked with water since you need it to purify water, so you don't get sick.

If you are in a cold environment, the shelter is likely going to be your most important survival component. Fire will also be important for staying warm and melting snow for water.

In the desert, you are going to be most concerned about water and then shelter or fire for more water and keeping warm during cold nights.

Jungle environments are going to need fire to dry clothes, purify water and keep predators away. The shelter is often second to keep you dry from a nearly guaranteed rain.

Lastly, in a wooded environment, you can generally prioritize in any order. Woodlands are often the most ideal for bushcraft skills. You can often find water easily, and the woods provide plenty of firewood and shelter sources. There is also an abundance of animal and plant sources for food.

Let's look at each of the four survival areas to see what makes them important.

WATER

Water is often at the top of the survival priority list. Serious and possibly fatal dehydration can occur in as little as a day or two. In a hot desert environment, you can even have fatal effects in less than 24 hours. Even mild dehydration symptoms can cause you to have reduced mental faculties leading to poor decisions that could lead to fatal consequences.

Water is of less concern if you are in a location with a lot of snow. If you have to, you can eat snow, but need to make sure it doesn't drop your core body temperature. Eating snow isn't as dangerous if you are active with traveling or working. However, you need to be eating snow while sitting at camp at night it can be dangerous.

If possible, you should always melt snow. If you absolutely can't melt snow, you should at least let is melt in your mouth or try to melt it using the outside of your body's heat, like a container filled with snow in your jacket pocket. This can help prevent the risk of dropping your core temperature.

SHELTER

You should never underestimate how important it is to have a good shelter. The number one cause of death is exposure; either too cold, wet and/or windy environments. Even in a temperate climate, these three aspects can be dangerous or fatal. In a cold and/or windy environment, you need to be insulated on at least three of four sides. Your shelter should be designed to reflect and trap heat from a fire. In a warmer climate, you may only need to block rain with a roof and have a raised platform to keep you off the ground.

FIRE

Fire is obviously something of great importance. Fire serves many purposes including purifying water, cooking food, drying wet clothes, keeping you warm, keeping away predators, keeping away insects, signaling for rescue, providing light, making tools and even improving the mood.

For these reasons, you should at least have two methods of starting a fire in case one either fails or isn't suitable to the situation. If you carry a fire starting method, you should also know a few methods of making fire without a fire starter. You never want to be dependent on technology.

FOOD

Food is often the last on the priority list, especially when it comes to short-term survival. However, it can become more essential if you are looking at long-term survival after a disaster. A healthy individual can live for weeks and possibly even months without food. Exposure and thirst are much more important in the short term.

When it comes to food in the wilderness, you can't be too particular. However, bushcraft skills do allow you to have a delicious meal. The main thing is to make sure you don't spend more calories getting your food than you do eating it. This would cause a calorie deficit, and this can be fatal if continued. The more energy you use, the more water and calories you lose.

Food is going to be more important in a colder climate than a warmer climate. A body burns extra calories in order to maintain a core body temperature. While food is often the bottom of the survival priority list, but it can become the top priority in a cold environment.

Another thing to be careful of is something known as "rabbit starvation." When you eat protein without fat, you can get sick with diarrhea in about a week. The body is not designed to handle this diet; you need fat. If this happens, even after you eat lean meat, you are still going to feel hungry. Along with diarrhea, you may also get a headache and feel uncomfortable; you can die within a few weeks if this continues.

What is available for food will vary depending on your location. If you can, you should learn about native plants in your area. Avoid eating anything you don't recognize. If you absolutely need to eat a plant, but don't recognize it then you can do a poisonous taste test.

To do this test, get some juice from the unknown plant's leaf and rub it on your lip. Wait a few minutes to see if you feel numb or zingy on your lip. If nothing happens then take a small piece with juices and put it in your mouth. Hold it in your mouth for a minute or two. Once again, if you don't feel anything then chew it up and let it sit a little longer. If still nothing happens then swallow what's in your mouth.

Wait at least eight hours to see if you feel sick or any discomfort. If nothing happens then you are likely safe to eat the plant, but keep your first meal small. If after another eight hours you have no side effects then you can increase the amount you eat. Some plants only become poisonous in large amounts.

Carefully consider the area around you and observe it to determine which four survival basics you need to prioritize. Failing to prioritize properly can lead to dangerous consequences. You also need to pay attention to how your body feels and meet its needs.

When You'll Need These Skills

Nobody wants a disaster to happen. However, accidents and injury can happen at any time. Even if you go on a camping trip prepared with the necessary gears, you can lose your equipment or find yourself without a valuable tool. You may find yourself in the wilds longer than your supplies would last or you even find yourself lost completely. There are a number of ways things can go wrong.

Just learning the basic bushcraft skills can give you the knowledge you need to stay alive in the wilderness. However, knowledge will only do you good if you practice.

You need to hone your bushcraft skills, so they become second nature. Just because you read about something doesn't mean you are going to know how to use it until you actually try it. Let's look at the specific steps you can take to hone your bushcraft skills.

PREPARATION STEPS

As you can tell, there is a lot to learn when it comes to bushcraft skills. To truly master bushcraft skills it will take several years or longer. However, if you are just getting started, there are a few skills you can learn quickly to get you started. There are also few basic skills such as making cordage and batoning branches that can have multiple uses throughout bushcraft skills.

For those just starting out learning bushcraft skills and techniques, it is best to start with the skills that are easier to learn. These are the skills you can learn within hours and further develop them as you practice.

Some of the basics you can choose to start with including the following:

☐ Batoning wood

☐ Carving basic tools

☐ Lashing basic structures

☐ Knot tying

☐ Basic fire starting

If you already have some basic knowledge of wilderness survival skills then you can work on more intermediate skills such as the following:

☐ Food foraging

☐ Primitive fire building without matches or lighters

☐ Shelter building

☐ Basic snares and trapping

☐ Water purification

Lastly, if you have a good working knowledge of survival or outdoor living then you can work on the most advanced bushcraft skills including the following:

☐ Making rope and cordage

☐ Advanced structure building

☐ Advanced foraging and trapping

☐ Land navigation

☐ Animal tracking

All of these skills can be learned with a few basic learning areas. Let's consider ten of the most common learning areas that can get you started on learning bushcraft skills.

10 Learning Areas for Bushcraft Skills

Bushcraft learning can often be broken down into ten basic categories. You can add more to each of these areas as you gain skills and find out what is needed in your specific location. Look at each of these areas to determine what order you want to practice your bushcraft skills.

ROPE CRAFT

Anyone who has spent time outdoors knows that rope craft and knot tying is one of the most important survival skills you can learn. Whether it be tying shelter knots or trying animal snares, rope craft can be very important. However, this craft can be taken a step further by knowing how to make rope.

Knife and Axe Use

Perhaps one of the most important and only tools used in bushcraft is the bladed tool. Knives and axes are vital to surviving and know the proper care and use of them is just as important as the survival basics. You need to learn not only how to use a bladed tool, but also how to sharpen, repair and make one if needed.

FIRE CRAFT

A lighter and matches make it possible for nearly anyone to start a fire, but do you know what to do if you don't have these? Would you know what material to look for in the wilderness if you needed help starting a fire? Fire craft is about more than just starting a fire. It is also about starting and maintaining a fire in any condition through a variety of techniques.

FORAGING

You won't survive in the wilderness with just fish and animals, especially if you don't have a fire to cook them on or the proper cooking techniques. This means foraging is also very important. You should research plants in your area and determine which are okay to eat and which are poisonous. In addition, you should research which are the best at providing your carbs and protein per ounce so you can preserve energy while foraging.

SHELTER BUILDING

Building a shelter is about more than just staying dry; it provides safety from various animals. Staying dry is more important than comfort since it will prevent you from getting sick and helps to keep all your gears dry and usable. You need to learn more than just how to build a basic shelter; you need to learn how to build them from a variety of materials, so you're never without the importance of a shelter.

FISHING

Learning to fish with a variety of methods and tools will ensure you have a good high protein and fat food source. You'll also want to learn how to make hooks and lures or any other methods of fishing based on the materials available to you.

TRACKING

There are many reasons to learn the art of tracking. It allows you to catch dinner, but it also allows you to be more aware of your surroundings. You can know what animals are likely to frequent your campsite. Recognizing tracks allow you to know where to set snares, get water and places to avoid.

HUNTING

It is pretty obvious that you need to hunt effectively in order to get food. However, you also need to know how to hunt with a variety of tools. Trapping and setting snares for small game are the best ways to learn to survive off the land.

In addition to knowing how to set snares and trap, you also need to know how to handle a trap once you've caught something. You'll also want to learn butchering and skinning techniques so you can actually eat what you catch.

CARVING

There are many wooden utensils you use every day without thinking about them. Wood knife handles, wooden spoons and a variety of other tools; this means knowing how to carve wood into a valuable tool is important. Learn how to work with wood from start to finish, and you can make just about anything in the wilderness.

CONTAINER CONSTRUCTION

Being able to construct containers can be a valuable and often overlooked skill. Your food, tools and clothes need to be kept in good and dry condition, and this is done by having a proper storage container for them. Learn to work with a variety of materials including leather, metal, wood and even tire rubber so you can make nearly any container you need.

When it comes to learning the basics of bushcraft, your focus should be on avoiding the mistakes of wilderness survival. There are eight common mistakes made when it comes to wilderness survival. Let's consider each of these so that you can avoid them.

8 Common Mistakes to Avoid

1. NO SHELTER

This is actually a double mistake that will cost your dearly in wilderness survival. The first is not having a proper shelter and the second is not having the right knowledge to build a shelter out of the materials around you. When discussing people who have gotten lost and died in the wilderness the term that often comes up is exposure (to sun, rain, cold, wind, etc.). No matter what form of exposure takes place, it is likely as a result of improper or lack of proper shelter.

2. Lack of Navigation Tools

If you go into the wilderness without at least a map and a compass you are asking for trouble. Anyone who has spent time outdoor knows that within seconds even an experienced individual can lose the sense of direction around thick trees or bushes.

In today's modern society people are becoming more reliant on GPS. Though you should never rely solely on GPS alone because technology can sometimes go bad and that is why you should always carry a map and a compass and know how to use them properly.

3. Lack of Knowledge

Not everyone who suddenly finds themselves in wilderness often doesn't have the necessary knowledge to survive.

Even if you aren't able to practice your skills before something happens; having general knowledge will still give you an advantage over others. Therefore, take the time to learn the basics of survival.

4. MISCALCULATING RISK

Survival situations often don't start off as such; like a day hike or a fishing trip. It is only when things go wrong that you are suddenly faced with life and death survival situation. Therefore, you want to plan for the unexpected. Before you set off into the woods, take the time to go through all contingencies.

5. IMPROPER CLOTHING

The general rule in the wilderness is to dress one layer warmer than you feel you need. You can always take layers off and save them for later. However, when you leave an article of clothing behind there is nothing you can do. It is best to wear clothes that retain warmth even when wet. You should also have a shell jacket and pants in case of rain or snow.

6. DRINKABLE WATER

Everyone knows the body doesn't last long without proper hydration. There are many waterborne organisms that will cause severe diarrhea and vomiting; which further increases dehydration and reduces your ability to survive in other areas. Therefore, it is important to know many ways of getting purified water, so you don't fall victim to dehydration.

7. Lack of a Signal Plan

In any survival situation, being able to signal for help is the key. There are a variety of tools dedicated to emergency signaling at most outdoor gear stores. However, what if you don't have those fancy store bought signals? It is a great idea to learn how you can signal with just the materials and tools you have at your disposal.

8. Lack of Fire

Fire is linked to many things in wilderness survival. Fire can keep you warm, provide protection, a signal for help and purify water. Therefore, I cannot stress enough that you need to know how to make a fire in a variety of ways based on the materials presented to you.

To help you avoid these common mistakes and have an easier time learning bushcraft skills be sure to consider some of the following tips and tricks.

Tips and Tricks for Bushcraft Survival

As you can tell, there are many skills and knowledge subsets that go into being more self-reliant in the wilderness when it comes to bushcraft survival. The more knowledge you have when you start off learning bushcraft survival, the better chance you have at being successful. So let's consider some tips and tricks that can give you a great head start.

SHELTER TIPS

If you carry a tarp with you, keep the guy lines tied onto the corners so you can avoid extra time and hassle when you need a shelter quick.

You don't want to pack a lot of gear when learning the art of bushcraft. However, when you are first getting started consider bringing a couple of contractor-sized garbage bags with you. You can fill these with leaves and use them as an insulated mattress.

Perhaps one of the most valuable additions you can make if you're going to carry any gear in a bushcraft situation is the blue tarp. Even if you just use it as ground cover, it will protect you from heat loss. In a pinch, you can even use the tarp as a quick and effective shelter.

FIRE TIPS

Always have two methods of fire starting. If you happen to always carry a lighter or matches with you, make sure the second method is the knowledge of how to start a fire with wilderness materials. If you don't carry a fire starter with you, then make sure you have knowledge of several ways to start a fire. You should never be caught outdoors without a way to start a fire.

While in the wilderness during the day, keep an eye out for dry tinder. You have to find a good source of tinder especially if you know you're going to be outside overnight, pick it up or at least remember where you can find it.

Food and Water Tips

If you're heading outside, always pack a protein bar or two. This way if something happens you at least have a reserve supply of food until you can trap or fish some more.

When heading outdoors, it can always be good to carry a method to purify water. Even if you have the knowledge to find and purify water, you don't want to be without water for too long.

Some of these tips you won't need once you become a master at bushcraft. However, for anyone new just starting out, it is better to be prepared, rather than head out into the wilderness and find you severely lack the basic knowledge of survival.

A lot of bushcraft is about learning what works for you and while it takes a little time to find out what actually works for you create a backup plan till you master your skills.

When preparing to learn bushcraft skills, the first thing you need to consider is your gear. Many are under the impression that you need a bunch of gears to survive in the wild, but this isn't exactly the truth.

Knowledge vs. Gear

When it comes to knowledge and gear, there is a big debate in the bushcraft community. Some believe that the true art of bushcraft is about going into the wilderness armed only with knowledge and a knife, everything else gets left at home. On the other hand, there are those who believe bushcraft is about being prepared and bringing the necessary gear to ensure survival in the wilderness.

Whatever your view is, take a moment to consider the pros and cons of each to see which meets your outdoor priorities.

GEAR

Pros	Cons
⍰ Increased capabilities	⍰ More expensive
⍰ Easier work	⍰ Many only work for specific situations
⍰ Saves time and energy	⍰ Can break

KNOWLEDGE

Pros	Cons
⍰ Can't become lost or stolen	⍰ Some tasks can't be done without gear
⍰ Always available and easily rebuilt	⍰ Work is more difficult without gear
⍰ Can adapt to any situation	⍰ More time intensive without gear
⍰	⍰

MIDDLE GROUND

Obviously, the best option is to take the middle ground. You want gear to help make things easier, but you also don't want to rely on your gear alone. Without knowledge and skills, gear alone won't help you survive in the woods.

However, when you talk to some practicers of bushcraft, they will tell you that there is no such thing as bushcraft equipment. You either practice bushcraft or you use tools; there is no middle ground. The heart of bushcraft is studying nature and resources you can use from nature.

Some bushcraft skills are pretty obvious and don't require a lot of knowledge to use. However, there are other resources that will require more extensive knowledge or greater skill levels. In between, there is plenty to learn. Given the amount of subjects and their depth, bushcraft can be a little overwhelming for new individuals to learn.

Therefore, the best thing is to ignore the majority of so call bushcraft supplies that are nothing more than glorified camping equipment (trust me I have seen it all). Rather just get a few basic supplies as a backup as you learn your skills and knowledge. Then as you gain in your bushcraft skill strength, you can continue to reduce the amount of gear your carry in your kit.

TOOLS YOU WILL NEED

Since you are interested in bushcraft(Since you are reading this book, I have to assume that you are), chances are you've already spent some time outdoors camping or hiking. Therefore, it stands to reason that you already have appropriate outdoor clothes. If this is the case you are pretty much properly equipped for bushcraft learning.

If you are a true beginner to bushcraft and the outdoors, then you may need just a few extra items. You might want to start with a little backup food, a fire starter and maybe even a sleeping bag until you improve and get comfortable with your skills. Still, there isn't much you need to pack when learning bushcraft skills.

Perhaps the most important thing you need when learning bushcraft skills is a basic cutting tool. This is what nearly all bushcraft masters always have with them. However, there are also some other bladed instruments you can choose to bring with you. Let's take a moment to consider various bladed instruments that you can bring with you as you learn bushcraft skills.

At the base level, bushcraft is about going into the wilderness and surviving with nothing but the clothes on your back, an edged tool and the knowledge in your head. Nearly all bushcraft skills and projects require you to use some form of bladed tool to make the process easier. Let's look at some of the options you can have available to you.

KNIFE

The fixed blade knife is the most popular among those who practice bushcraft. This knife needs to be well-built, so it gives you years of service without breaking or rusting. These knives are typically suited to medium field duty like carving or whittling, skinning game, making snares and traps, preparing food or cutting down branches no bigger than the size of your wrist.

Hatchet or Tomahawk

These are small, versatile and non-precision wood cutting tools that are better suited to heavy tasks like chopping wood or downing trees, splitting logs, butching larger animals, digging or hammering posts and stakes.

MACHETE

A machete is a combination large knife and an ax. It can be used for a number of lighter jobs but also has a long heavy blade that makes a great cutting tool. They are good for clearing brush, cutting large branches, chopping smaller logs or digging. There are some machetes that are sickle hooked to remove vines.

SAW

This specialized bushcraft tool is used to cut wood branches for a variety of projects including building platforms, huts, storage buildings, camp chairs, table, Bush ladders or just to cut firewood.

Backpack or Not?

Obviously, a master of bushcraft isn't going to have many supplies and may not require a backpack to carry them. However, if you are just starting out and carrying some backup supplies, you may want to bring a backpack, so it is easier to move and carry your supplies.

If you are going to bring a backpack you want to make sure it is comfortable and matches your body type. You also want to choose one that is waterproof and has multiple compartments to organize all your gears.

There are few other tools you may decide to carry such as cooking pots, matches, compass, blue tarp and few others, but I decided not to list them here as I will discuss those in later chapters and then you can decide if you need them in your backpack or not (I also included a list of suggested back pack items for your first trip, which you will find at the end of this book).

Whether you are going to stay within a narrowly defined area or if you are going to move around often to accommodate changes or hunting or other activities, then a backpack can be an essential piece of gear for a bushcraft practicer.

Now let's take a look at the individual skills in the four survival basic areas that you need to learn in order to survive in the wilderness.

How to Find & Treat Water

Two the four survival areas involve finding adequate food and water, then preparing or purifying it properly. As we've already discussed, water is one of the most important survival basics and often needs to be addressed before other survival basics. The first thing you need to do is know how to search for water. Let's consider four signs you need to look for to find water sources in the wilderness.

4 Signs to Look for When Searching for Water

VEGETATION

Take a look at the landscape around you; do you see small trees, bushes or clusters of tall grass. If this vegetation grows in a line, then there is likely an underground stream. Dig a small hole at the base of a group of plants to confirm this.

Insects and Birds

You can often find water if you directly follow insects and birds. Bees particularly require fresh water to survive and often build their hives just a few miles from fresh water. If you find a beehive, then you should immediately look around for signs of water.

Mosquitoes are quite irritating, but they often breed in pools of standing water. The Mason fly also uses mud to build and as a result often frequents moist soil and underground springs.

When it comes to birds, wild pigeons are a reliable indicator of water. These birds will feed on grain all day and then seek water at dusk. Pigeons fly low and swift when headed towards a watering hole.

On the other hand, if they are flying from tree to tree then they are likely returning from a watering hole. When they are weighed down with water, they slow down and use more caution when flying. When you carefully observe insects and birds, you can likely find a water source.

ANIMAL TRACKS

Grazing animals need to drink in the morning and evening hours in order to digest their grass. If you find a hoof print, look downhill to find a potential water source. In addition to finding more tracks to follow, you can also keep an eye out for snapped twigs, scat, scraped bark and other signs left by large animals. Often the path to a watering hole is well used and clear of obstacles. A carefully trained eye will be able to see signs of wear that indicates a trail.

TERRAIN

The terrain around you can also be an excellent roadmap to finding water. Water flows downward due to gravity, so you will stand a better chance of finding water by seeking low ground. If you walk parallel to a mountain, you will likely find an outlet of fresh water or at least a dry stream bed.

While a dry stream isn't of any use to you in a survival situation, you may still be able to access water beneath the surface. The best stream bed to investigate further is one with dark green vegetation. However, any vegetation is always a good sign. Look for dark patches of earth or dampness, the outer side of a bend or natural depressions to find the best places to dig.

You can access underground water sources by digging a seep. This is a hole that is two to three feet in diameter and at least a foot deep. Once you dig your seep, groundwater should slowly seep into the hole. Line the bottom with rocks to prevent sediment from getting stirred up by the water.

Fresh groundwater is safe for you to drink, but it is always best to sterilize it first. Also if you leave your new hole unattended, then wildlife might share and your water is no longer purified.

Dehydration causes devastating effects that no one wants to experience. It is important that you learn to find water since it is essential to your survival in both the short and long term. However, there are plenty of sources of water in the wilderness. Let's take a look at the many sources of your water.

Wilderness Sources of Water

When it comes to short-term survival, you may be okay in the water department, but even if you are going to be in the wilds for over a day you should start looking for water sources.

In a long term survival situation, this is even more important. Thankfully, there are plenty of sources of water in the wilderness. Let's look at the many ways you can get water in the wilderness.

RAINWATER

While rain may seem like a great thing to someone suffering from dehydration, you should be aware that rainwater is actually not safe to drink because of pollutants in the air. However, you can always purify this water to make it safe to drink. We'll discuss several of these purifying techniques later in this chapter.

Collecting rainwater itself is as easy as simply setting up containers. The containers should be placed in unobscured areas in order to get direct rainwater and not water that is dripping off plants.

PLANT SOURCES

If you can find a green plant, you can find water to harvest. There are several ways to get water from plants. Just make sure you don't choose a poisonous plant. Otherwise, plants can be an excellent source of fresh water supply.

As a general rule, you should avoid any milky or colored sap or plant juice. The only two exceptions to this rule are the juice from the American Barrel Cactus and coconut milk. There are several ways to harvest water from plant sources, let's take a look at them individually.

TRANSPIRATION BAG

Plants use photosynthesis to take up water, and during this transpiration, process water is released into the air. In order to capture this water, you can place a clear plastic bag over the end of a leafy branch and secure with a cord or rope. Within a few hours, you can have several ounces of water. Again you want to purify this water for safety before drinking.

Directly from Leaves

Plants are good at surviving droughts. There are some plants that have leaves which serve as a natural cup. These plants use the leaf to funnel rainwater towards their stem. Look for plants that have leaves growing directly from the stem and you can often find several pints of water this way. Typically these types of plants are only found in tropical areas.

Water from Tree Trunks

With a tree, xylem help transports water from the roots to the leave in a vertical matter. You can collect water from tree trunks the same way sap is collected from maple trees. To harvest water with this method, you will need a strong, tubular stick about the same diameter of your thumb or a hollowed out piece of bamboo.

Sharpen the tube at one end and gently tap into the trunk with it at a 70-degree angle. It only needs to go in about a few inches. Set up a collection reservoir below the tube to catch any dripping water. While a bucket is ideal, you can also use a plastic bag or a large leaf. The collection can take awhile with this method, but the water you get will be safe to drink.

Water from Roots

The roots of plants do contain water, but harvesting this water is a labor intensive task. You first need to start by cutting a large root and stripping the bark. Then you need to use rocks to mash the root into a pulp, producing droplets of water and pressing the pulp into a container for collection.

The roots of some trees produce more water than others: the blood wood, the water tree, and the desert oak; all of which are located in Australia carry a lot of water. While this method will get you water, you shouldn't use it if you are going to expend too much energy.

Water from Bamboo Plants

Bamboo plants can be a great source of water in the wilderness since it is stored in the cavities between the joints. You want to find bamboo plants that are the most yellow since these will often have more water. After choosing a piece of bamboo, tap it and listen for a low thud; this indicates that it's not hollow.

Next find a section with water, cut a notch just above the lowest joint and collect the water when it runs out of the bamboo. This water is safe to drink directly, but you should still purify it just to be on the safe side.

Water from Vines

Vines can be another source of water, but you also need to be careful with your selection. Some vines have a milky sap that is poisonous. If you don't see any milky substance, you can proceed by cutting a deep notch in the top of the vine.

Next, cut off the tip of the vine to let water flow and work your way up the vine until there is no more water flowing.

It is important to start at the topmost part of the vine first. Otherwise, it will draw all the water back to the base of the plant. Some vines will yield a decent amount of water. Don't allow the vines to touch your lips as some can be very irritating and itchy.

Water from Cacti

Most cacti have some level of fluid content, but not all will yield safe drinking fluid. Some cacti are poisonous, such as the giant Saguaro cactus found in California, Arizona, and Mexico.

However, the Barrel cactus is the exception to the avoid milky fluid rule. This plant can yield about a liter of drinkable sticky milky sap. The best way to get fluid is to cut off a piece to expose the inner flesh, then either cut chunks out of the center to mash or suck the moisture out of the cacti.

Always be careful when handling cacti since you will have difficulty getting the spines out of your skin once they are stuck and sores can become septic.

Other Sources of Water

Water from Snow and Ice

As we've already discussed, in cold and snow conditions you can still suffer from dehydration just as easily as in a dry and hot climate. The snow and ice around you is an obvious source of water, but you always want to melt it before consuming.

If you simply eat snow or ice, it will lower your core body temperature, increasing your risk of hypothermia and dehydration. Eating ice can also lead to blistering and sores in the mouth and on the lips.

Melting ice is more efficient since it requires less energy than snow melting. However, melting snow may be your only option. If this is the case and you don't have a fire, fill your drink bottle with snow and tuck it under your clothes while moving so your body temperature will slowly melt it.

If you need water immediately and can't wait for a whole bottle to melt, then you can melt a small amount of snow quickly by rubbing it between your palms and letting it drip into your mouth.

Lastly, in a few situations, you will be melting down sea ice for water. If this happens, you want to use old, blue sea ice with smooth edges since it will be lower in salt content. If you don't have old sea ice, you can still distil sea water to get rid of the salt and make it safe to drink.

Water from Condensation

In a survival situation, humans are at risk of fluid loss when breathing, talking, sweating and digesting. However, most living things lose fluid in this same way, and bushcraft allows you to increase your chances of survival by using this knowledge. If you heat up anything that contains water, the water will evaporate.

The difficult part is catching it, so it condenses and provides you with a decent amount of drinking water. Thankfully, there are a number of ways that you can collect condensation.

When you collect condensation, it is evaporated water that removes impurities. Therefore, this process provides you with fresh, drinkable water from a variety of sources.

We will discuss distillation process later when talking about purifying water, but they all follow the same principle of using heat to evaporate water from a substance.

The trick is catching the water vapor and cooling it before moving it into a container for drinking. There are many options depending on where you are and what materials are available to you.

Obviously, the easiest option is a large plastic bag as long as the fire is low in intensity. You would make a tripod from three five to six-foot sticks and position it over the top of the fire.

The plastic bag is placed at the top of the tripod, creating a cap over the fire. Fold the opening edge of the plastic back up inside the bag to create a catchment for the condensed water. Place the container of liquid on the fire and wait for the water to gather.

Another option is to use a large, clean piece of fabric rather than a plastic bag. Once the fabric is saturated, you can replace it with a dry piece. Allow the saturated piece of cool for a minute or two and then wring the water out and place in a container to drink.

You can also use a large flat leaf such as a banana leaf. Fashion it into a cone shape and pin it with a stick to catch the water vapor that evaporates. A smaller tripod can be placed in the water container which would suspend the leaf cone above the fire. You can use bamboo split in half lengthwise laid around the fire to catch the condensed water as it drips off the edge of the leaf.

Now that you know how to find and harvest water let's consider how you can purify it and make it safe to drink.

4 Ways to Filter Water

After you've found and harvested water, it is often full of debris such as leaves, dirt, insects and other foreign materials. This makes the water unsuitable for drinking and needed to be filtered to remove impurities.

Water filtration is not the same as sterilization. While filtration will get rid of debris in the water, making it taste and smell better; it won't remove bacteria. This means that you will still need to boil or distil it to sterilize the water unless you are sure the water is fresh.

There are four main ways to filter water with items readily available in the wild. The basic idea of water filtration is to remove debris of all sizes by passing the water through different sizes of filters. Some good filtration materials include sand, dried grass, pebbles and charcoal.

The type of filter that you make depends on the resources you have available and how you want to use the filter. There are four variations on the filter above, but with a little resourcefulness that are limitless options to filtering water.

Perhaps the best filtration material is charcoal. The carbon in charcoal will attract impurities and improve the smell and taste of the water. If you light a fire or come across charcoal, then you should take some with you. Charcoal serves a number of uses in the wilderness beyond water filtration. It sparks easily for fires, can be used as camouflage and can be used to reduce glares off your cheeks.

1. Quick and Simple Filter

Combine a few of these in layers of smallest to largest in a sock or your pant leg/shirt sleeve and tie at the end. Pass the water through the layers of material and catch the clean water as it comes out the end. You should run a couple of cups through the filter first to remove any dust or charcoal powder.

2. GYPSY FILTER

This option for filtering water can be a good option if you can't boil water. Start by digging a hole beside a pool of water, about one foot away and deeper than the water level in the pool.

Let it sit for awhile and eventually the water filters through the dirt and sand to fill your well. The water will be murky at first so you should scoop out the water and let it fill up again. This time the water will be clearer and good to drink.

3. CHARCOAL STRAW FILTER

Sometimes you're the only source of water is too shallow to fill a cup. This is when you want a straw filter. You will need to find a long and hollow item like a straw. In a pinch, you can use a piece of frame from a backpack or a hollow stalk from a dried plant or reed. Stuff it with a wad of dried grass and push it down with a stick. Place some crushed charcoal and then follow with sand. Finally, place another piece of dried grass to pack it down; not too firmly, but enough to hold the material in place. You can now drink water directly through the straw.

4. BAMBOO FILTERS

You can cut a section of thick bamboo so that both ends are hollow. Pack one end with a wad of grass and then fill it to 3/4 with sand for an excellent water filter.

PURIFYING WATER USING SOLAR

Once you have filtered your water, you should ideally still distil it to remove bacteria. There are a few ways you can do this in the wild. The obvious method is to boil the water, but if this option isn't available to you, there are two other solar options available.

SOLAR STILL

Perhaps the best way to extract water directly from the ground is with a solar still. To build one you'll need the following supplies:

☐ Collection device

☐ Large sheet of plastic

ROCKS

☐ Long straw

After gathering your items, you need to choose a sunny spot to dig a wide hole. At the base of the hole, you want to dig a fitted spot that is large enough to set the collection device.

If you have leafy, green vegetation in the area; place it in the hole around the opening of your collection device to enhance the collection rate since it will draw moisture from the plants along with the air.

Next, you need to place a sheet of plastic over the top of the hole and use the rocks to secure it. Place a small stone at the center of the plastic to create a low joint point that is just above the top of the bucket for condensation to collect into the collection device by dripping.

Even in a desert climate, a solar still has the potential to collect up to a quart a day. You can use a long straw or another tubing to access water without disturbing the still.

Solar Disinfection of Water

This is another simple and easy way to distil water. With this option, you use the sun's UV rays to break down the bacteria in the water. If you have a container for water, you simply seal the water container and lay the bottles out in the open, exposed to the sun for a period of time according to the chart below depending on weather conditions.

Weather Conditions	Minimum Treatment Duration
Sunny	Six Hours
50% Clouds	Six Hours
50 to 100% Clouds	Two Days
Continuous Rain	Collect Rainwater

Now that you have a safe source of water let's take a look at what you can eat while in the wilderness.

How to Find & Prepare Food

If you need food fast, foraging is your best option. So let's take a look at how you can do proper foraging in the wilderness.

Foraging for Food

Even in favorable conditions, it can be difficult to feed yourself off the land. Therefore, food is one of the more important bushcraft skills you can learn. To start, you want to have good knowledge of local plants and animals. The first step in bushcraft skills is to learn how to forage for edible plants. When you can forage for edible plants, it gives you a quick source of food as you take the time to set up snares or go fishing for other food sources.

5 Ways to Foraging

1. Roots and Tubers

These can be found in the soil under a vine or stalk of a plant. These are nutritious, but often require some form of boiling or cooking. Potatoes, yams, and onions are some commonly known roots and tubers.

2. GRASSES

Many kinds of grass have young whitish tips that are edible and often quite tasty. You can easily eat these raw.

3. Seeds and Nuts

Many plants have edible seeds and nuts that provide you a good source of nutrition. If you notice a bitter or acidic taste when eating a seed or nut, they are likely not safe to eat. You can often make nuts and seeds safe to eat by soaking them in water for 12 hours or boiling.

4. Fruit and Berries

A lot of fruits and berries in the wild can be harmful. Often you should avoid any fruit that is red in color. Unless you can be sure a fruit or berry is safe to eat (like strawberry or raspberry), you should avoid them.

5. LEAVES

Many plants have edible leaves either raw or boiled. The leaves of watercress and nettles can be very tasty; both of these grow near freshwater streams. Again, avoid eating leaves that have a strong bitter taste.

5 Things to Avoid

Knowing what to avoid in foraging is just as important as knowing what you can eat. However, there are always exceptions, and this is why you want to know as much about local plants as possible. Consider some general guidelines as to what you should avoid.

1. BAD SMELLS

If a plant or fruit has an odd odor, then it is probably best to avoid eating it. Our sense of smell often warns us against dangerous items.

2. Taste of Almond

This flavor is often associated with the presence of a prussic acid which is toxic. You can sometimes remove this by boiling. If, after boiling or soaking, you can't taste the almond bitterness then it is likely safe to eat.

Make sure you safely discard of the water that you boil the plant in where it won't contaminate any humans or animals.

3. Bitterness or Acidity

Any plant that tastes bitter or "hot" needs to be avoided. This is often a sign that a plant will make you sick.

4. RED COLOR

Anything red is likely to be dangerous to humans, leaves or fruit included. There are some exceptions in the case of strawberries and apples, but unless you are absolutely sure a plant or fruit is safe you should avoid it.

5. FUNGUS

There are some tasty and edible mushrooms. However, it can really be difficult to determine safe from toxic mushrooms. Therefore, unless you are positive a mushroom is safe to eat; just avoid them all.

Plants vs. Animals

When it comes to calories, animals are obviously the most efficient food source for survival. Plants offer additional nutrients, flavor and supporting calories; but often don't have enough calories to sustain you. Most leafy plants only have thirty to fifty calories per ounce of weight.

In comparison, small freshwater fish have 200 calories, and a single squirrel can be 500 calories. The numbers are substantially higher for large game. Nuts do offer 150 to 200 calories per ounce.

If you are going to eat plants, let us consider fifteen edible plants that you are most likely to encounter in the wilderness.

15 EDIBLE PLANTS

When possible do a Google search to see each of these plants so you can identify them later in the woods.

1. Cattail or Bulrushes

You can eat the leaves, stems, flowers, pollen and roots. They can be harvested anytime. They are a tall, straight plant with large, cigar-shaped and colored flower head. They usually have long, linear leaves about one to two cm wide.

Typically grown in swamps, wetlands, and many open wet areas. They can also be useful for making baskets and mats, as a torch and as packing material.

2. CHICORY

You can eat the leaves, roots, and flowers. Often harvested in spring or fall since the summer heat makes the leaves bitter. Featured by purplish-blue flowers about two to four cm in diameter. It is a branching plant that is scraggly looking and often grows alone in open areas along roadsides, grassy areas and fields.

3. COLTSFOOT

You can eat the flowers and young leaves. The flowers are harvested in early spring and the leaves in late spring. It is a dandelion-like flower growing on a slim stalk. The flowers bloom and die before the leaves appear.

The leaves are basal, hoof-shaped, waxy and smooth on the surface with white hairs on the underside. It typically grows in ditches and open areas along the edges of forests or places prone to spring flooding. The plant has low levels of toxicity and should be limited to no more than 6g a day. The roots are potentially toxic.

4. DANDELION

You can eat the leaves, flowers, and roots. It can be harvested anytime. It features a bright yellow flower with deeply notched, hairless and basal leaves.

5. ELDERBERRY

The berries and flowers are edible. The flowers are harvested in spring and the berries in late summer. This shrub is distinguished by smooth, gray-brown bark and umbrella-shaped flower clusters.

The berries are purple-black and grow in clusters. Aside from the flowers and berries, all other parts of the plant are poisonous.

6. GARLIC MUSTARD

The flowers, leaves, roots and seeds of this plant are edible. The leaves can be harvested anytime but will become bitter when the summer heat arrives. The roots can be harvested before and after the flower stalks develop. The seeds are harvested in the fall.

The plant is identified by broad leaves that are heart or kidney shaped and have course, rounded teeth. The flowers are tiny and white with four petals shaped like a cross.

The plant often has the odor of onion or garlic. It typically grows in the open, in disturbed forests, wooded areas, ditches, swamps, and roadsides.

7. HORSETAIL

All above ground parts of this plant can be eating when cooked or dried. It can be harvested after spores are released, and leaves grow. The plant doesn't have flowers but has a brown stem that releases spores in the early spring. The joined stems and tube-shaped leaves branch at each joint. It grows nearly anywhere.

8. LAMBS QUARTERS

The leaves, shoots, flowers and seeds of this plant are edible. Spring is the best time to harvest this plant, but anytime during the summer will also work. The seeds can be harvested in the fall.

It is identified by tiny green flowers growing in clusters on spikes. The leaves are diamond or goosefoot shaped, light green on top and white underneath. The seeds do contain saponins, which are mildly toxic so you should only eat them in small quantities. You can cook the leaves to remove the oxalic acid.

9. MILK THISTLE

You can eat the roots, stems of young flowers, leaves, flower buds and seeds. The flowers are harvested from June to August. The plant is distinguished by growing up to one meter with a branched stem that takes two years to grow.

The leaves are prickly and uniquely shaped with a light purple flower. When harvesting this plant make sure you have hand protection.

10. PURSLANE

You can eat the leaves, stems, and flower buds. The plant can be harvested all summer long. It is identified by a thick, red steam with small, thick and spoon-shaped green leaves. The plant grows anywhere with poor soil including wastelands, fields, roadsides and cracks in the road or sidewalk.

11. RASPBERRY

The berries are the only edible part of this plant. The berries are harvested in late summer when they become red or black and ripe. The plant can be identified by its spade-shaped, toothed leaves that are light green on top and light greenish silver on the bottom.

12. ROSE HIPS

The fruit is the only edible part of this plant. The fruit is harvested in the fall after the flower petals have fallen off the plant. This is your typical rose bush, but the rose hips are bright red or reddish-orange round fruit.

13. STINGING NETTLE

The edible parts are the leaves, stems, and roots. You have to harvest the plant before it flowers. It is identified by narrow, fine-toothed leaves that are covered in tiny, sharp and stinging hairs.

The flowers grow from leaf axils and are tiny greenish blooms. You should never eat this plant raw and have exposed skin covered when harvesting.

14. Wild Carrot or Queen Anne's Lace

The edible parts of this plant are roots, leaves, flowers, and seeds. You can harvest when the flowers bloom and when they go to seed. The roots of most young plants are edible. It is identified by the large, flat clusters of white flowers that look like lace.

It generally has a single dark purple flower in the center. The roots often smell just like carrots. The leaves are feathery, broad and flat. It typically grows in fields, meadows, roadsides and waste areas. Make sure you know your local plants well since it can often look like Poison Hemlock or Fool's Parsley. Since Poison Hemlock is deadly, be absolutely sure before eating.

15. WILD VIOLET

The flowers and leaves are edible. It is harvested early spring into summer. It is a low-growing plant with heart-shaped leaves and blue-violet flowers on each stalk. It typically grows along riverbanks, streambeds, in thickets and woods.

As you can tell there are sources of food all around us. A regular bushcraft project you can do is to practice identifying and gathering appropriate food items. However, if you find yourself in a long term survival situation you want to move on to more substantial food sources. An excellent choice is to learn to catch fish if you are near a water source.

4 Ways to Catch Fish

One of the more popular wilderness survival food options is fish. Fishing is a very vast topic, and there is a lot to cover. There are some small fishing kits and tools you can carry with you in the wilderness to give you an advantage, and you can also build fishing tools once you are in the wild.

However, when it comes to bushcraft skills, the focus is again on doing things without added tools or gear. So we are going to look at some basic methods for catching fish if you find yourself out in the wilderness without anything to help you fish.

1. BARE HAND FISHING

While there are other simple ways of fishing without tools, this is still a valuable thing to learn. The best way is to sneak up from behind the fish. Fish often face upstream, so slowly place your hands in the water and slowly close in on the fish and then quickly grab with both hands.

Be aware that with this method you are going to keep your hands and feet in the water for extended periods of time. This can lead to hypothermia if you're not careful. This is especially true if you don't have medical aid nearby or supplies in a true survival situation.

2. TROTLINE METHOD

This option is good for fast-running waters or a fishing location that isn't near to a bank. Put two solid stakes a few feet apart and tie them with a cord. Make sure the cord and stakes are below the water surface. Then you want to tie one or more shorter cords to the main one and place a hook and bait on the loose end. You can place as many smaller cords as you want, but they should be placed far enough apart to avoid entanglements.

3. SPEARFISHING

Perhaps one of the first methods you should learn and practice for bushcraft survival is spearfishing. A spear is easy to make and can be effective at catching fish. However, spearfishing will only work in shallow water. But if the water is above your waistline, then you might want to consider other options such as snorkeling.

4. FISH TRAPS

Fishing traps offer a big advantage to survival situations since they require very little energy and preserving energy is important in survival situations. One of the easiest fish traps to make is a basket trap in the shape of a funnel that you can make by putting several sticks together into the shape of a cone.

3 HELPFUL FISH CATCHING TIPS

There are a lot of ways to catch fish. Consider the following three tips to help make it a little easier for you.

Know where to look for fish. Even if you don't see them; check under rocks, ledges, and crevasses.

If you are partially hidden behind a rock, you can often wait for the fish to come to you. Sometimes this can be the better option if you're trying to save energy.

No matter what fishing method you choose to use or what tools you may have; keep in mind the most important tip: the less aggressive you are underwater, the more fish you'll catch.

Now if you're in an area where this isn't viable fishing options or you need additional sources of food you want to consider the small game. Let's look at some snare and trap options that can help you get small game.

How to Make Snares and Traps

When it comes to small game, there are plenty of methods for catching food in the wilderness. However, bushcraft skills center around knowing how to make snares and traps. Snaring and trapping works by either choking, crushing, hanging or entangling the animal, so it is safe for you to approach it.

As with most survival techniques, the best traps are those that are the simplest. You can get the upper hand by having knowledge and understanding of animal habits so you can position your traps in the best locations for a higher chance of catching an animal. Let's take a look at the basics first.

Most snares and traps will use a combination of two basic methods. All of these methods can be done without any equipment or gear.

THE SIMPLE SNARE

This snare is exactly what it sounds like, a basic trap construction that uses minimal equipment and materials. These simple snares can be constructed within minutes and are the fastest to set up, but the harder part is knowing where to set your trap.

Looking for animal runs and paths can be a challenge. The simple snare uses a wire or rope with a sliding noose on one end while the another end is firmly anchored down to something solid. You want to try and size your snare to the type of animal that you want to trap.

It is important to note that a simple snare won't hold an animal for long, so you want to check the traps often to prevent predators from taking off with the food. The best use for a simple snare is for squirrel. When you set up several of these, you are guaranteed to have some success in trapping food.

THE SPRING SNARE

This is an ideal method of trapping animals because it secures trapped animals out of reach of other predators. First, you need to set up the snare along an animal trail or pathway such as one leading to a waterhole. Second, you need to find a young and live tree branch near to the trail that can be flexible and able to bend down to ground level.

Once you've found the right location, make a trigger stick with 'V' notches as shown in the picture below. Part B needs to be longer and driven securely into the ground.

Next, tie a length of cordage to Part B. Now pull the young tree down and make sure there is enough tension to spring back up when released. Tie a length of cord to the young tree and the other end to the smaller trigger peg or Part A. While holding the young tree down places the trigger peg into the 'V' of Part B.

Lastly, make a noose with the rest of the cordage. You want to use a slipknot so if the animal tries to force its head through; the loop will tighten. To help the noose maintain shape, push some slender sticks into the ground and drape the loop over them.

As the animal walks into the snare and through the noose it will pull the peg loose and release the tension causing the young tree to spring up and snare the animal.

These are the two simplest methods to trap animals and can be a good source of food. However, one you have your food you need to prepare and cook it. Let's look first at how you can prepare your fish and meat.

Preparing Fish and Meat

Gutting and Preparing Fish

The first step after catching your fish is to gut and prepare them. There are several different ways to process fish. The most essential part is to remove all the innards.

Make a slit along the belly and then push your finger into the cavity and peel out the innards. Rinse the cavity well with fresh water. You must remove all innards before cooking since any traces can ruin the meat and potentially make you sick.

Most bushcraft practicers prefer to cook the fish whole and then remove the skin and head. This will save you a lot of time and effort. However, if you want to fillet a larger fish into portions, then you will need to scale the fish before cooking.

You should also remove the fins and gills before cooking. The area also needs to be washed out and cleaned before cooking.

Properly preparing a fish is an important skill and one you should practice. Fresh fish needs to be handled carefully to prevent it from falling apart when you start to cook it. In the next segment, we're going to discuss methods of cooking. For now, let's take a look at how you can prepare game.

Butcher and Prepare Meat

Once you have learned the art of trapping your food, you need to know the safe procedures for butchering and preparing food. The most important thing to remember is that you need to process and prepare your meat almost immediately, especially in warmer weather to prevent spoilage. The food is edible for only a matter of hours.

You should field dress any game you catch at the spot you catch it. To remove the entrails, start at the tail/anus and cut into the belly until you reach the chest cavity. When slicing into the belly make sure you don't puncture the entrails. You should bury all entrails right away, so you don't attract any predators and insects.

Next, you should leave the skin on until you are ready to process the meat for cooking. Cut off the head and feet then spread the cavity open and prop with a stick to wipe the cavity out completely.

The meat should cool as quickly as possible; so don't leave any entrails in the animal any longer than necessary. You can increase your knowledge in this area, by becoming proficient through hands-on experience processing wild game.

Next, skinning an animal is easier when hanging it by the hind legs. Make a slit in the skin along the backside of the hind legs. Start to peel the skin by starting at the hind legs and work forward, essentially pulling the skin inside out towards the head. If you know how you can keep the skin for curing, otherwise you should bury it as soon as possible.

For a small game, you can simply cut through the joints to remove the limbs. While the legs of the small games don't have much meat, you can save them and boil them in water to make a protein-rich broth.

Now you can make an incision on either side of the backbone and right through the ribs to free the breast meat. If you want to dry some meat, place it to one side on a drying rack. Otherwise, skewer the animal and place it over your fire to cook.

When you know how to properly prepare food in the wild you can prevent food-borne illness, but also you will be making sure you are eating well. The last part of eating in the wilderness is being able to cook your food without any tools or utensils.

4 PRIMITIVE COOKING METHODS

Another important bushcraft skill is the knowledge of wilderness cooking techniques. You may not always have cooking utensils or equipment, so it is important to know what alternate methods you can use to prepare your fish and meat that you've caught. Let's consider these seven primitive cooking methods.

Ash Cooking or Cooking Over Coals

With this method, you simply create a bed of coals. Spread the coals and make an indentation in the coals to lay the fish or meat. If you have wrapped the fish or steak in anything, you can cover it with coals. Cook time will depend on the size of the fish or steak.

BROILING RACK

For this method, you can find some flexible branches to fold into a tennis racket shape. Use branches to add crossbars to create a sort of broiling rack. You can use another branch to tie the fish or meat down. The goal is to keep the food at a distance from the flame or build a bed of coals to slow roast the food.

PLANK COOKING

This is another cooking option when you don't have any pots or pans. You need to use a non-poisonous wood and also avoid woods that have strong tasting resin such as conifers. Ideal woods to use would be poplar, cedar, and oak.

You can split the wood with an ax or baton it with a knife. You can tie the food to the board or notch some pegs to place on the board to keep your food from slipping.

Once the food is secured to the plant, you want to place it near the coals. If you can hold your hand for a few seconds before it starts to feel uncomfortable, then your fish is the right distance from the coals.

Hold the plant horizontal by placing another piece of wood behind it. If you have a firewall, you can also place the wood vertically and lean it against the firewall. The fish will slowly start to brown.

Hot Rock and Steam Pit

With this option, you dig a hole and line it with rocks. Build a fire inside the hole, so it heats up the rocks. Wrap the food in non-poisonous leaves and place at the bottom of the pit. Cover the food with some of the soil you dug out and allow it to sit a few hours. The food will come out cooked and moist.

Now you know how to find and prepare your food. Another important survival skill you need to learn with bushcraft is building a shelter. Let's take a look at this area next.

How to Build or Find Bushcraft Shelters

Shelter is another one of the main survival basics. A shelter will become even more important when you are in an extreme environment, like harsh cold or harsh heat. However, a shelter can be just as important in a jungle or woodland environment.

The type of shelter that meets your needs will depend largely on your environment. Do you need it to block the wind or shade the sun? Does it need to contain heat or be rain proof? Do you need a hunting advantage or protection from animals? These factors and others will determine the shelter you need to build; then you need to use the materials available to you to help build it.

In addition to being a protection from the elements, you want a shelter that is comfortable enough to allow you to get good sleep.

Sleep is essential in outdoor living and survival. Sleep deprivation can lead to mental loss and clarity issues; this causes you to make poor decisions which are very dangerous in a wilderness survival situation. When it comes to shelter building, there are the 5 W rules that you want to consider.

5 W RULE CHECKLIST FOR SHELTER BUILDING

There are five important factors when it comes to choosing your bushcraft shelter location. This is referred to as the 5 W's checklist by professional bushcraft practicers. Let's consider these five to see what you need to consider when building your shelter.

WOOD

Obviously, you would have a difficult time building a shelter without wood unless you are using a natural formation or other materials. However, you'll also need wood for a fire and other tools. So before you set up your shelter make sure you have a good supply of wood nearby.

WATER

Obviously, you want to make sure your shelter is located near a known water source. If there isn't a water source nearby, you will likely need to abandon your shelter and move to a place with water. So it is better that you just set up your shelter in an area where there is water.

WEATHER

Weather is something serious that you want to consider. I've said it before; exposure is the number one cause of death in the wilderness. The weather around you will affect the shelter you need to build. For example, in a hot environment, you won't need much shelter other than shade while in the cold you will want a shelter that is completely insulated to block the wind and hold in heat.

If you are in an area where rain is highly likely, then spend extra time making sure your shelter is waterproof. You don't want to be caught unaware and have to rush to improve your shelter.

WIDOW MAKERS

This refers to dead trees that can fall suddenly. Don't ever place your shelter under or near a large dead tree that hasn't fallen yet. However, this category can also be expanded to include objects such as rocks or any other inanimate objects that have a potential to fall and injure or even kill you. Take a moment to look around the area where you are planning your shelter to make sure it is a safe spot.

WILLIES

This section refers to any critters that can make it difficult for you; such as snakes, spiders, scorpions, bees, hornets, bats, ants, mosquitoes, ticks, *etc.* A shelter needs to protect you from these threats. If there are a lot of creatures on the ground, such as you find in a jungle; consider a raised platform. In general, avoid making your shelter in an area that looks like a habitat for any potentially poisonous animals.

Remember always to go through this 5 W checklist when choosing a location for your shelter to make sure you are avoiding any issues down the road.

CHOOSING THE RIGHT SHELTER LOCATION

Another thing that goes into choosing your location is how long you plan to use your shelter. For an immediate, short-term shelter you may be able to just use a naturally occurring shelter. If you are going to need a long-term shelter, then there are several other things you need to consider when choosing where to build your shelter.

You want to be close to water, but not so close that you are more prone to insects. Also, the noise of running water can mask the sounds of other dangerous animals. You don't want your shelter on a hilltop where you are prone to wind and lightning. If cold weather is an issue, then you want to stay away from valley bottoms since cold air sinks.

You also want to stay away from areas that have a risk of flooding. Mountainous areas are prone to flash flooding. So stay away from low areas or dry stream beds.

Wind chill is an important factor since it dramatically increases body heat loss. If the climate is cold, you also don't want to sleep directly on the ground. You can use pine needles, foliage, fluffy bark or plant matter to make an insulated raised platform.

It is also important to make sure your shelter has ventilation to let air cycle through. Without ventilation, you are at risk of carbon dioxide poisoning, musky humid air and/or trapped smoke. You ideally should have two holes in your shelter; one near the bottom and one near the top. If you have a snow shelter, then you want to check the holes regularly to make sure they haven't become clogged with snow.

No matter where you choose to set up a shelter, you always want to be aware of the wildlife around you. Use a stick to prod areas and stir up any creatures that may be hiding in bushes or under fallen logs. Store any food you have high up in a tree to avoid attracting unwanted critters.

You should also avoid building your shelter anywhere you see chalky white substances on foliage. This is the appearance of mold and it can spread to your shelter, causing negative health effects. You also want to avoid a tree that contains lacy leaves since this means insects have probably infested the area.

Now that you have a decent idea of where to build you shelters, let's consider what type of shelter you should build.

Types of Shelters

First, take a look at this simple table to help you see some of the basic options and which might work for you. Then we'll take a look at them in a depth.

Type	**# Occupants**	**Time to Build**	**Difficulty**
Simple Frame	1 to 2	Less than an hour	Easy
Body Heat Shelter	1 to 2	Less than an hour	Easy
Open Shelter/Lean-To	1 to 4	3-5 Hours	Moderate
A-Frame	1 to 3	3-5 Hours	Moderate

Teepee	1 to 2	3-5 Hours	Moderate
Subterranean Shelter	1 to 8	1 hour to a week	Easy to Difficult
Log Cabin	1 to 4	Weeks	Difficult

NATURAL SHORT TERM SHELTERS

If you are only going to need a short term shelter until help arrives or you are able to hike back to civilization, then there are some natural shelters that can work for you. Be on the lookout for these good natural shelter options if you need a quick and easy short-term shelter option.

TREES

A large fallen tree can make a good shelter, especially evergreens. Break, cut or move any branches that are directly under the fallen tree and move them to the side, so you have room to move inside.

Make sure the tree is secure and won't fall on you. If the base of the tree is still sitting on a stump, this gives you some great space to camp under as long as it isn't at risk of collapsing. If it is safe, this will be similar to building an A-frame shelter.

Trees provide a dense, low canopy of leaves and branches that can effectively provide a good shelter in mild environments. Another option is a large tree that is hollow inside.

CAVES

This can be an excellent shelter option. However, approach all caves with caution since a lot of animals like to live in them as well. As you approach, make loud noises, so you don't startle a predator when you enter. Throw sticks and stones into the cave from a distance and listen for any noises. In addition to large predators, you may also find snakes and rats prefer caves as well. If you have a good fire, they will typically leave, but make sure you leave an escape route open for them.

Some caves will even have a fresh water source, especially if the cave runs deep into the hillside. The cave may have a stream, or the water may seep in through the rocks.

Before you shelter in a cave, always check for rockfall. Some caves regularly drop chunks of rock, and others don't. Look on the floor and in sharp cracks in the ceiling for any signs of recently broken rocks, if you see anything then don't use this cave as a shelter.

When you light a fire in a cave, heat it up slowly. If the temperatures change quickly in a cave, then rocks might crack and fall. It is best to put the fire in the cave rather than near the entrance.

This will give you more warmth, and the smoke will travel up to the ceiling and then out the entrance. If you place your fire at the entrance of the cave, a lot of heat is lost, and the smoke may blow in where you are. If you are in a shallow cave, the fire should be placed at the back.

SUBTERRANEAN SHELTER

Lastly, if you need a short-term shelter consider a subterranean shelter. This can be as simple as a hollowed out mound of earth that gives you a warm place to sleep. The ideal location is the root base of fallen trees since the roots will provide structure and prevent the shelter from caving in on you.

However, there may be times and situations where you are going to need a long-term shelter. If this is the case, then there are a few options to provide you with durable, long-term shelters with no tools and limited supplies. Let's look at these options one at a time.

BODY HEAT/SIMPLE/DEBRIS SHELTER

This is a simple shelter that is easy to build and useful for short-term or unexpected situations but can also be adapted into a long-term shelter if needed. With this shelter, you gather debris from the ground like dirt, leaves, and twigs to create a mound and use few large sticks to build a frame around it. Then clear a hole just big enough to crawl into it and cover the opening to block air flow and limit the amount of open space. You body heat will become trapped in the shelter so you can stay warm overnight.

This type of shelter offers two main advantages:

1. It is comfortable and warm, even without a fire.
2. It is easy to set up and only takes a short amount of time since you don't have to cut wood.

There are also two disadvantages to consider when thinking about this type of shelter:

1. Unless you strengthen the structure, it won't last too long.
2. If the wind is too strong, the dry leaves may blow away. But you can add medium weight branches over the top to keep them in place.

Open Shelter or Lean-To

This shelter is the next step up and gives added protection from the elements such as wind and rain. You can build these as large as your resources allow so you can potentially accommodate a lot of people. Construction can take two to five hours depending on the materials available and how big of a shelter you want to build.

Start by looking for fallen trees that have branches low enough to support the topmost point of the lean-to, called the ridgepole. If you only find one tree you can use it as a ridgepole, but if you find two you can lay a sturdy branch between them.

Next, gather about five to six poles to lean against the ridgepole at about a 45-60 degree angle. Make sure you leave enough space to create a comfortable fit for you underneath the shelter. This serves as a grid. To create the grid frame, attach five to six poles across the frame. Weave flexible boughs between the poles at right angles and then use bark or leafy branches to thatch the roof; start at the bottom and work your way up to the top.

If needed you can add additional walls for more protection in the same manner. If you happen to have a tarp with you, you can hang it from the opening to be a curtain.

This type of shelter has three main advantages:

1. Easy to build.
2. More durable than an A-Frame shelter.
3. With a wall in the front as a heat reflector, it can be pretty warm.

There are also three disadvantages to consider when deciding whether to build this shelter:

1. To build a really warm and good lean-to, you will need to do a lot of work.
2. You will need a lot of natural materials to make it waterproof.
3. While they can be expandable to fit more people, they are generally best for just one person.

A-FRAME SHELTER

This shelter is constructed similar to a lean-to shelter. The only difference is that the ridgepole starts on the ground and then extends up into a tree and is lashed at a height that allows enough space to sit underneath. The two sides are then constructed to create the A-Frame shape, providing good protection from the weather. You can locate a fire pit near the opening for added warmth.

There are four advantages to this type of shelter:

1. It is the easiest shelter to build.
2. You don't need geometric measurements to put the materials together.
3. You can build it pretty much anywhere.
4. With enough natural materials, you can make it waterproof.

There are three disadvantages to consider with this type of shelter:

1. If you are planning to stay long-term, this may not be the best type of shelter.
2. This shelter won't work in damp areas or places with wet ground.
3. It is hard to build a fire near the entrance.

While these are the most common types of bushcraft shelters, there are also a few others that work in individual situations that you may want to consider. Let's see what these options are, some of them will be good for long term survival.

SOD SHELTER

The sod shelter or house is a very labor and time intensive option that is only practical to build when you need a long term place to live. You are also going to need a good grass source in order to build this option.

First, you want to cut and pull up turf rectangles from the ground. They should be about one to three inches deep and six to eighteen inches wide. You then stack them to build the walls. You lay them like bricks with an alternating fashion.

Build one wall taller than the opposite one and connect them at an angle so you can attach a pitched roof to deflect the rain. The roof should be built with wooden poles and then covered with leaves or sod in a shingled (similar to typical roofing shingles) fashion.

TEEPEE

With this option, you gather five to ten wooden poles and tie them together at the top with them spreading out at the base. The more poles you use for the structure, the stronger it will be; especially against the wind.

If you have a tarp, you can wrap it around the poles. Otherwise, you can use branches, leaves, mud, moss, bark, grass, *etc.*

First, you need to weave bendable branches throughout the stacked poles in order to create a framework to place the debris on for the outer walls. Birch bark makes excellent panels when cut in large pieces.

A teepee is nice because it allows you to have a fire inside, as long as you leave a hole in the top for the smoke to escape. You can also cover the hole if you need to when it rains.

JUNGLE SHELTERS

When you are in a jungle environment, the most important thing is to be raised off the ground when you sleep. Anytime the ground is wet, or there are dangerous insects and reptiles on the ground then you should construct a raised bed.

Drying out regularly is also important. If you are wet, then skin ailments can develop. A fire can help keep you dry, but in a wet environment, you want a rain proof shelter. Often in a wet environment, you will find plants with large, broad leaves that make for excellent roofing materials.

To sleep off the ground, you will need to construct a swamp bed shelter. There are several ways to build this, but generally consists of making four pegs in a rectangle and then securing wooden poles on the pegs.

Lay wood poles across the supports and secure it with cordage if available. You can stack soft material such as plant matter or dirt to make the bed more comfortable.

SNOW SHELTER

If you find yourself in a cold climate or somewhere with snow, you are going to need to build a snow shelter. Snow is a great material that can provide both a windbreak and insulation.

Use heavy snow to build up walls and compact it. You can stack a lot of conifer or evergreen branches on top for a roof and then add a thin layer of snow for extra insulation. Check the roof regularly and remove any excess snowfall to prevent the roof from collapsing with heavy snow.

When digging for a snow shelter, make sure you use an implement, stick or with gloves. Never dig with your bare hands to prevent frostbite.

Now you have the basic knowledge you need to build a bushcraft shelter. The next step is to go out and practice building the various types of shelters. While some of these techniques can be practiced in your backyard, there's no substitute for going out in the wilderness and trying the real thing.

When practicing take note of how long it takes you and what methods work best for you. This way you will know what type of shelters you should construct in the event you find yourself in a situation when you really need one. Now let's move on to learning how to start a fire.

BUSHCRAFT FIRE MAKING TECHNIQUES

When you find yourself in a survival situation, the most important thing that can make the difference between life and death is the ability to start a fire. While a well-prepared person may always have a fire starting tool on them; there is always the chance this tool will get wet, lost, used up or malfunction. If this is the case, bushcraft fire skills give you the knowledge you need to create a fire naturally without any modern fire starting tools. Let's consider five methods of starting a fire without tools.

Five Ways to Make Fire without Tools

Flint and Steel

This is the best bushcraft fire starting method. All that you need is a flint and a piece of carbon steel, such as your bushcraft knife. Flint can be found in the wilderness if you don't have one on you, we'll discuss finding flint in the wild later. With a flint and steel, you will be able to start a fire in pretty much any condition.

The process takes five steps:

1. The friction caused by striking steel against flint will cause sparks.
2. Strike the steel against the flint with a loose wrist.
3. Have tinder ready to ignite as the sparks fly.
4. Gently blow on the sparks that land on the tinder until it ignites.
5. Make sure there is kindling nearby and ready to add to the tinder as it catches on fire.

This method of bushcraft fire starting is relatively easy to learn. It requires minimal physical strength and energy, plus it can be done with two simple items.

FIRE PLOW

This is a form of fire starting that uses only natural materials. It is essentially an optimized version of rubbing two sticks together. Basically, a base board is one, and the plow stick is the other. Start by finding a flat piece of hardwood at least a couple inches thick to serve as the baseboard.

Here is a 9 Step Process to Start a Fire:

1. On the flat side of the baseboard, cut a straight line down the center.
2. Hollow out the line to create a shallow, thin groove that is about 1/4 inch wide.

3. Find a softwood stick about one inch thick. Use your knife to carve the end of the stick to a rounded point, with the tip being small enough to fit into the groove on the baseboard.
4. To cause friction to generate heat to ignite a fire, rub the two crafted parts together.
5. Applying continuous force, rub the stick through the groove from one end to the other.
6. You will need quite a bit of strength to create the friction required to form a spark.
7. As the sticks rub, wood will slowly shave off the baseboard.
8. The friction will create heat and tiny embers, which ignite the wood shavings.
9. Have kindling ready and gently blow on the sparks within the tinder to ignite a flame.

While the fire plow is a simple and versatile fire starting method, it does require a bit of physical force and energy. This means it may not be the best option for someone who is weakened by starvation or exposure.

HAND DRILL

A hand drill fire starting technique is easier than a bow drill but does require more energy, patience, and skill to complete. The only supplies you will need are a drill and a fireboard. The stick being used for the drill is spun between your hands to generate friction to create fire embers.

This is a Seven Step Process:

1. Find a softwood stick and carve one side down to a rounded point.
2. Cut a small hole in a softwood fireboard, about an inch from the edge of the board.
3. Cut a V-shaped notch connecting the hole and the edge of the board, with the point of the V connecting at the hole.
4. Fill the V-shaped notch with tinder and position the point of the drill into the hole.
5. Place both palms flat on either side of the drill stick. Press your hands in firmly and rub back and forth. Apply downward pressure as you spin the drill to give added friction.
6. Continue to spin the drill until smoke and embers form.

7. Once the embers start to make the tinder smolder, blow gently on them until a flame develops.

This method is valuable to learn since it can be put together quickly in an emergency. But sure to practice this technique often since it can be difficult to master.

BOW DRILL

This method of fire starting is a little more complicated to construct than the previous methods. However, after the construction; a bow drill requires less exertion to create a usable ember.

This is a Ten Step Process:

1. First, you want to find a piece of hardwood, rock or bone that has a divot or shallow depression. This serves as the socket that the drill will rest in during the fire making process.
2. Next, find a straight stick that will work as your drill. The drill needs to be hardwood and about 3/4 inch thick. One end needs to be blunt, and the other end needs to be chiseled to a rounded point.
3. Make a flat piece of softwood about one inch thick to use as a fireboard. Cut a tiny hole in the board about one inch from the side and just big enough for the tip of the drill to fit.
4. Now cut a triangular notch connecting the hole to the side of the board, with the point of the triangle connecting to the hole.
5. Next, find a bendable green stick to use as the bow.
6. Tightly tie a piece of sinew or paracord to the bent ends of the bow.
7. Place the tinder into the triangular notch in the fireboard since this is where sparks will form.
8. Place your foot firmly on the fireboard to hold it in place. Loop the bowstring around the drill and place the point of the drill into the hole in the fireboard. Grab your socket from the first step and place it in the depression in the socket on top of the drill to hold it in place firmly.
9. Pull the bow back and forth rapidly. This will drive the drill into the wood and create friction.
10. As the embers start to drop into the tinder next, gently blow until it ignites.

While this is obviously more complicated to build, it will create fire faster with less effort. Once you make your tools for this fire starting method, you should be able to reuse them multiple times in case you need to start another fire or move to another location.

PUMP FIRE DRILL

This is the most difficult fire starting method to construct. This is because it is more mechanically complicated than the other methods. However, as with the bow drill method, this one requires nearly no effort to start a fire once it has been constructed.

This is a Seven Step Process:

1. Find a round piece of hardwood and cut a small hole through the center. Using sinew, cordage or paracord attach a sharp rock or arrow to the end of a hardwood stick.
2. Drill a hole through a rock at the same width as the stick and push the stick through the hole with the arrow sitting about one to two inches under the rock. The rock should have a snug fit without moving.
3. Find a slightly curved piece of hardwood and drill a hole through the center the same width as the stick.
4. Insert the curved piece of wood onto the stick about an inch above the rock, so the bow in the wood faces upward. It should have a loose fit so the device can easily rotate inside of it.
5. Etch two notches on the ends of each side of the bowed wood so that a string can be tied around easily. Connect the edge of each side of the bowed wood to the tip of the stick with a piece of sinew, cordage or paracord, in the direction away from the arrow.
6. Grab the bow wood to use as a handle. Press the arrow to the hole in the firewood and spin the device to wind it up and prepare it.
7. Allow the device to unwind. As the arrow is driven into the fireboard, embers will form.

This bushcraft fire skill requires almost no effort to create a fire and is perfect for those who have little strength. However, it does take some time to build the drill so you should practice building it ahead of time.

What is the Best Method?

In the perfect world, you would be able to master all five of these fire starting methods and be ready to build a fire under any circumstance.

However, not everyone has the time and ability to master all of these bushcraft skills. Rather you should pick one or two methods and practice them until you know the process backward and forward. Picking two methods comes down to considering the following factors:

☐ What resources are most likely to be available in your area.

☐ How much time are you likely to have to make a fire.

☐ How much energy are you able to expend to make a fire.

You can also try out all five methods in your backyard to see which works best for you. Then once you have singled out one or two methods, you can focus on practicing them regularly.

The time you spend improving your fire starting bushcraft skills can be invaluable if you ever find yourself needing a fire, but having no tools on hand. However, knowing how to start a fire won't do you any good unless you can get the proper items you need to start a fire. First, let's consider how you can find a source of flint in the wild.

Gathering Flint in the Wilderness

Flint or chert is a type of sedimentary rock that serves a variety of uses. At one time it was used in tools like knives and spears. Today, Flint is typically used to create sparks for a fire when struck against steel. If you don't carry flint with you, then you need to know how you can find flint in the wilderness.

It isn't that hard to identify flint, but it is often only found where there was once an ocean. It is most common in the South East and Mid West. Quartz is another type of rock that can be used like flint to start a fire. Agate in the Mid West can also be used similar to flint.

LOCATING FLINT

First, you want to choose the right area to search. While it may seem difficult to find flint, you generally just need to choose the right area to look. In some areas, you can find flint lying all over the ground. This is because flint is a hard, durable rock that is so resistant to weathering that they remain intact long after other rocks have weathered into the soil.

(Flint can come in varieties of colors, most typical are this type of shiney black type)

Ideally, you want to search along freshwater shores or riverbeds. Flint is durable and resistant to chemicals, so it often collects in the remaining soils as carbonate rock erodes. Rocks like limestone will erode, and the fine soil is carried downstream, small pebble deposits of flint will collect along the shores.

You can also try other locations where there is a large variety of rocks, such as construction sites or gravel roads. Rocks are often harvested from riverbeds for construction so you might be surprised where you find flint pebbles.

KNOW LOCAL HISTORY

If you live near an area that was once home to Native American tribes, then you might stand a decent chance of locating flint fragments. Flint was the preferred choice for creating tools and weapons. If you find an arrowhead or sharp rock near an old tribal area then you know flint is in the area.

Look in Larger Rocks

Flint often forms within larger pieces of chalk or limestone. So in addition to looking for pieces of flint, be sure to look for larger rocks that may contain pieces of flint. Bust these rocks open to see what you might find.

Look for discolorations in limestone. Usually, flint or chert will be a slightly darker shade than the limestone. You can break these pieces out with your knife and collect the flint.

SPOTTING FLINT PROPERTIES

It is important to pay attention to the color of the rock. Flint often appears black or dark gray. This is the only physical difference between flint and chert. Chert doesn't have an identifying color, but it typically appears in a combination of shades depending on what other minerals are present.

Shades of maroon, tan, yellow, white or occasionally deep blue are common among chert. Sometimes these colors will also form bands along the surface.

You should also learn to identify other types of quartz that can be used as flint. Such as carnelian, agate, bloodstone, jade and chalcedony.

Surrounding rocks can influence the appearance of flint. For example, when flint is buried in chalk a white patina or film can form over the flint.

You also want to look for various shapes of flint. Flint can occur in natural nodules or as a fragment that has been worked into a shape. Nodules can appear in various smooth, rounded shapes within chalk or limestone. When you find flint within a chalk bed, it is common to see an imprint of shells cast into the surface.

Keep an eye out for rocks that have been split like broken glass. Flint fractures differently from crystals. When pieces come apart, they tend to look like glass shards with curves and sharper edges.

You also want to keep an eye out for flint that has been worked into a shape. You are able to control the way flint splits easier than other rocks, which is why people used to make tools and weapons from flint. Flint may sometimes have edges that seem to have been chipped away or have a point, showing they have been used as a tool.

Look for rocks with a glossy surface. Flint often has a natural, glassy luster. If the flint was just broken, the luster may seem dull and feel somewhat waxy. You can often rub away or sand the cortex to reveal more surface luster.

Test to see how hard the stone is. If you have a glass bottle, you can scratch it with the sharp edge of the flint. Just be careful when striking glass with a rock, it can be a good idea to wear gloves to protect your hands.

Strike the stone with your knife. If sparks fly after several attempts, then you likely have a piece of flint. If the rock doesn't have sharp edges, you will need to create one to test for sparks. Make sure the stone is dry before striking, since a damp stone will not produce sparks no matter how hard you try.

In addition to locating flint, you also need to have tinder to start a fire. There are four good options for tinder. Let's consider what options you can choose from.

FOUR FIRE TINDER OPTIONS

CEDAR SHAVINGS

If you collect the stringy shavings from the bark of a cedar tree, these make excellent natural fire tinder. Rub the tinder between your hands quickly to create a small fluff ball of shavings that will light quickly.

Cotton Ball and Vaseline

This is a trick many people don't know about but has been around for awhile. If you have a first aid kit, coat some cotton balls in Vaseline. The Vaseline is flammable and will burn hot for longer than you would think.

CATTAILS

While these aren't available in all areas, the light feathery material inside cattails works just the same as burning paper or cotton. It is a quick and easy fire tinder option.

BIRCH BARK SHAVING

As with cedar, just shave some bark from a birch tree. The birch shavings are similar to paper. They are thin, light and easy to burn. They also peel off the tree easily.

This covers the survival basics of bushcraft skills. However, there are still some other areas of bushcraft you need to learn about before heading out into the wild to practice. The next area you should consider is first aid and safety while out in the wilderness.

Bushcraft Safety and First Aid

BUSHCRAFT KNIFE SAFETY

Bushcraft skills rely on the main tool you carry; the knife. However, you need to know how to properly handle the knife in order to use it safely. Perhaps the first thing that many experts will tell you is that everyone cuts themselves at some point.

While this may or may not be true, you can follow a few practices that will reduce your chance of a serious injury and possibly eliminate cutting accidents completely.

The important thing is to instill a safety first attitude. If you are expecting to cut yourself, you are going to be more cautious about how you use your knife. Know the dangers of using a knife when fatigued, in poor light or while distracted, and you will be able to use a knife safer and avoid serious injuries. This means you need to work slowly and concentrate on what you're doing. Don't cut corners for convenience if it is going to risk your safety. Combine this safety attitude with the following nine knife safety tips to improve your knife safety skills.

Sharp is Safer than Dull

For bushcraft beginners, this may seem like a paradox. You never want a knife that will slip off what you're cutting and go out of control.

For example, imagine carving into a tree branch. With a dull knife, you will need to exert a lot of force to cut into the wood since your knife is going to act more like a spoon. The more force you exert, the more likely your blade is going to slip. When you knife slips, it is more likely to be out of control.

Therefore, keep your knife sharp. Learn how to sharpen your blade and properly maintain the edge after use.

8 KNIFE SAFETY PRACTICES YOU MUST FOLLOW

1. Keep Your Eye on the Blood Bubble

What is the term "blood bubble"? I wondered this same thing when I learned this. The blood bubble is an invisible sphere that comes from the center of your body. The radius stretches to the tip of your blade when your arm is fully extended.

It acts as a "no fly zone" for your body. Your blade has the ability to reach anything or anyone within this radius. It is your job to monitor your personal blood bubble. When someone comes within this radius, stop using your knife. If you are walking near others using their knives, then display courtesy and common sense by avoiding their personal radius or blood bubble. Using this precaution will provide a safe environment, even when many people are using knives.

2. Don't Put Pressure on the Point

The purpose of a bushcraft knife is to slice. They often have a tiny finger guard, if any. The thought is that a finger guard makes it difficult to choke up on the knife and maintain certain grips.

However, this means that if you push on the point of the knife, it is likely your fingers will slide onto the blade. So avoid stabbing point first with a bushcraft knife. Instead use a safer technique that will get the same accomplishment.

3. Avoid Using Your Body as a Backstop

You should never have a reason to use your body as a backstop. If your knife slips, it will cut or stab you. Rather, put the wood you are carving against a tree, on a table, on the ground or any other area other than your body.

Also, make sure you position your hands in such a way that they are out of harm's way. Often people will use their body as a backstop without thinking of it, so always take the time to think things through first. Always consider where a knife will go in case it slips when you make your next cut.

4. Never Attempt to Catch a Falling Knife

If you lose your grip on the knife or you see it slips off a table, ignore your natural instinct to try and grab it.

Train yourself instead to jump clear of a falling blade. Even if you are sure you can grab it, avoid this temptation. Nothing is worth the pain of getting cut. If you need, you can always buy a new knife.

5. USE CAUTION WHEN SHEATHING

A lot of times people use caution when drawing their knives. Then they concentrate and work carefully with a knife. However, when returning the knife to its sheath, they relax and don't use as much caution. This is when your guard is down, and you aren't paying attention. This is human nature.

However, you should wait until your knife is safely in its sheath before you relax. Look at what you're doing and use caution.

6. Concentrate and Take Your Time

There are a lot of things that can lead to an accident: distractions, rain, mosquitoes, fatigue, cold and dehydration to name a few. When you are using a knife, do so with concentration and deliberation. Be mindful and notice your surroundings.

7. KEEP ELBOWS ON KNEES

The human body is remarkable at adapting to the environment. For example, the femoral artery is on the inside of the leg since most trauma and attacks are going to come from the outside of the leg. So if you're carving in a sitting position, never use your knife directly over your legs. This exposes your artery to injury if your knife slips. Rather rest your elbows on your knees to extend your knife and the material you're carving in front of your legs.

8. TWO STEP UNSHEATHING

Most knives require friction to stay in the sheath. A few sheaths have a locking mechanism that keeps the knife in place. When unsheathing a knife, you need to overcome the static friction or the locking mechanism. This requires more force than drawing it the rest of the way.

If you draw your knife in one motion, the large force you exert means the knife can easily become out of control. To avoid this, you should always unsheathe your knife in two steps.

For the first step, you want to overcome the static friction or the locking mechanism with your thumb. Place your thumb against the nub or button built-in to a sheath.

If you don't have one of these, then use the top edge of the leather or plastic above the spine of the knife. Your thumb should be flexed. Grip your handle and extend your thumb until straight. This draws your knife enough to overcome the friction.

For the second step, you should slowly draw the knife until it is completely clear of the sheath.

Always avoid grasping the sheath with your other hand when drawing a blade. The most likely time for a sheath to fail is during the drawing or sheathing of your knife. You don't want your hand around any part of the blade, even if a sheath is between you and the blade.

In addition to basic knife safety, you also want to have some first aid knowledge. While there is no rule that says you can't bring a first aid kid; a true bushcraft practicer doesn't bring any gear and this means you need to know how to treat ailments with natural plants around you. So let's look at some herbal foraging techniques.

HERBAL MEDICINE FORAGING

A first aid kit is always useful when it comes to survival. However, what do you do if you find yourself lost in the woods and injured without a first aid kit? Or in a long-term survival situation, how are you going to replenish your first aid kit?

How would you treat a common ailment rather than an emergency? One way to solve all of these issues is by learning herbal medicine foraging skills. This allows you to identify and harvest natural remedies from the plants that surround you. Let's look at twelve of the best herbs and the illnesses they treat.

12 Herbs That Works as Medicine

1. ECHINACEA

Also known as coneflower, this is a popular herb that is best for fighting colds. Aside from boosting the immune system, it is also a beautiful flower. These flowers grow naturally in Eastern and Central North America; often in moist to dry prairies and open wooded areas.

2. ASTRAGALUS

This is another good immune booster. However, it is also a good antimicrobial and antibacterial. It can be used for conditions such as colds, flu, diabetes, heart diseases and even the side effects of chemotherapy. The plant grows along the edges of woodlands in thickets and open woods as well as grasslands. It is a hardy plant that even survives North American winters.

3. LICORICE

While this plant is sweet, chewing on the root is the best remedy for a sore throat. You can also make it into a tincture or decoction. The plant grows in warm climates often in deep, rich sandy soil near a stream in the full sun.

4. CHAMOMILE

This is the best herb for stomach aches. You can make it into a tea and drink it along with honey for a calming effect. It typically grows along fence lines, roadsides and in sunny fields from Southern Canada to Northern US. It doesn't do well in hot, dry climates.

5. WOOD BETONY

This plant is in the mint family and is often used for stress and headaches. It can also be used to cover minor wounds and help relieve inflammation and soreness. It doesn't do well in harsh conditions. It typically grows in woodlands and around copses of trees, although it can sometimes be seen in open areas.

6. CALENDULA

This flower can be eaten directly or mixed into muffins. In addition, you can use it to soothe the skin and heal rashes. You can make it into a salve to easily use for skin irritations. It is typically found growing wild in open fields and prairies.

7. FEVERFEW

This is another herb that does well for headaches and can also help dilate blood vessels. It can be used as a tea or tincture. You can often find it growing along roadsides or the borders of wooded areas.

8. ELECAMPANE

This plant is a member of the sunflower and ragweed family. It can grow as high as five feet tall. The herb works to stimulate the digestive system and relieve congestion. It grows in pastures and along roadsides in wet, rocky ground.

9. HOREHOUND

This herb works great for coughs. However, it is extremely bitter and requires a lot of sugar or honey in order to make it easy to take. It is a hardy plant that grows well in full sun with just a little moisture. It is typically found along roadsides, in disturbed areas, and in and around fields.

10. VALERIAN

This herb root doesn't smell great, but it works in a decoction or tincture for relaxation because of its sedative properties. It grows best in moist soil in marshes or along river banks.

11. MARSHMALLOW

This herb works as a mucilage to coat the throat and stomach. It also has anti-inflammatory properties. It grows in sunny but cool climates along the edges of marshlands and on grassy banks near lakes and rivers.

12. COMFREY

Also known as bone knit, this herb works great for external wounds. You never want to take it internally since it can be toxic to the liver. However, you can mash the leaves and soak them in hot water for a few minutes then wrap them around a wounded area. It typically grows in damp, grassy places.

How to Use Medicinal Herbs

Now you know what type of herbs and plants can help common ailments and injuries. But how can you actually use these herbs? There are a few options when using these herbs; except for comfrey which is only used externally. Let's consider what options you have for making an herbal remedy.

TEA

Perhaps the most popular way to use medicinal herbs is to make them into a tea. You simply need to create an infusion by boiling water and adding the leaves to it. This is easy enough to do in the wilderness if you don't have any tools or utensils. Allow it to steep for about ten minutes, strain the herbs and use it as your drink.

DECOCTION

If you are using the roots and bark of a plant, you can make a decoction by adding a handful of the fried or fresh herb into a pot of water. Allow it to simmer after boiling and then strain the herb out; the decoction needs to be used within a day.

TINCTURE

To create tinctures, you can use alcohol in order to remove the oils from the herbs and preserve them. You should use dried herbs and enough alcohol to soak it about a quarter inch. Shake it every day for one month to eight weeks then strain and bottle. The regular dose of tincture is about thirty drops three times a day.

It can be a lot of work to use medicinal herbs, but in the long-term, if you need to have something to treat common illnesses and you don't have access to a first aid kit; this knowledge is important.

Learn about local herbs that grow in your area and try to identify them next time you're out on your hike in the woods.

Another important bushcraft skill you should learn is navigation. Any outdoor person will tell you not to venture outside unless you have at least a compass and ideally a map. Using these two items let's see what you need to know about navigation.

Navigation Skills you Need to Master

MAP READING SKILLS

When navigating through unfamiliar areas, a map is invaluable. Whether you used them to plan your route in advance or you're relying on them to get your through your expedition; a map is something you'll use every day. However, a map won't do you any good unless you know how to read it.

Map reading and land navigation are two vital bushcraft skills that you can apply to a number of situations.

MAP SCALE

First, you want to have a map with a useful scale. For example, a small-scale, highly detailed map won't do good if you are just driving through an area; but a less detailed map won't do good if you are on foot in the wilderness.

This is why you want to understand the scale bar. The scale bar shows the size at which a kilometer or mile is shown on the map and is often expressed as a ratio. For example, 1:50,000 means that each measure on the map will be 50,000 times smaller than the actual distance. This allows you to use the map to find distances.

CONTOUR LINES

Understanding how to read contour lines is also useful when it comes to map reading skills. Maps are 2D, so different heights are indicated by contour lines. Contour lines show the altitudes of the land and are recorded at regular intervals, often fifty feet. Each point on a contour line is hypothetically at the same height, indicated by a number. This tells you how high above sea level the terrain is.

Contour lines that are closer together represent a steeper gradient. However, you also have to consider the scale of the map, so you don't over or under estimate the gradient.

MAP KEY

The map key will explain any symbols used on the map. These can range from manmade to natural structures including types of land such as woodland, swamps or beaches. Some features won't be depicted to scale. For example roads, paths, and waterways often have a standard width that may not represent exact measurements.

MAP GRIDS

Maps have both vertical and horizontal grid lines that divide the map into squares. This division is either based on latitude and longitude or may be specific to the particular mapping authority. Grids help you determine distances more quickly since they are usually on a comprehensive scale.

ORIENTATING YOUR MAP

Keep in mind that the grids on a map don't exactly indicate north and south, although they may provide an indication of this. Also keep in mind that a compass points to magnetic north, not true north. Most maps also show magnetic north. These deviations can help you read your way across a landscape with a map since you can take note of your surroundings in relation to map features.

This is just the basics of map reading. You can always take courses to further your knowledge of map reading. Even if you know an area, consider taking a map with you so you can practice map reading and land navigation skills. Once you master these skills, you'll never have to worry about getting lost again.

How to Use a Compass

If you regularly spend time outdoors, a vital bushcraft skill to learn is how to use a compass. Before the popularity of GPS, a compass was frequently used for a variety of outdoor experiences. Every compass has at least four main directions.

Along with a good topographical map of the area, knowing how to use a compass will nearly guarantee you never get lost. Let's look at the basics of a compass and how you can use one for all your navigation needs.

Types of Compasses

There are actually two types of compasses that you can use. Let's take a look at the two types, so you can see which one works the best for your individual needs and likely situations.

BUTTON COMPASS

This is the typical type of compass that you find in cheaper survival kits. This work fine if you need to get a simple direction reading or figure out which direction is North. However, they won't be very helpful when it comes to navigating the woods if you don't know where you are or which direction you want to go.

ORIENTEERING COMPASS

Also known as a mountaineering compass, this compass has a magnetic needle that always points to magnetic north. It also has a dial or bezel that you need to rotate in order to find which way to go. There is a direction of travel arrow and also meridian lines that can be oriented to the maps longitudinal lines.

Within this category, there are prismatic and lensmatic compasses for more advanced outdoors individuals. These compasses have additional features. Some have phosphorous enhancements so that the directional arrows will glow in the dark.

Learning the Basics

LAYOUT

While there are different types of compasses, all of the have a magnetized needle that orients to the magnetic fields of Earth. The basic compasses features the following components that you need to know:

☐ Baseplate - the clear, plastic plate on which the compass is embedded.

☐ Direction of Travel Arrow - The arrow in the baseplate pointing away from the compass.

☐ Compass Housing - The clear, plastic circle that houses the magnetized compass needle.

☐ Degree Dial - The twistable dial surrounding the compass housing that displays all 360 degrees.

☐ Magnetic Needle - The needle spins within the compass housing.

☐ Orienting Arrow - The non-magnetic arrow within the compass housing.

☐ Orienting Lines - The lines within the compass housing that run parallel to the orienting arrow.

How to Hold a Compass

Place the compass flat on your palm, with you palm in front of your chest. This is the proper stance when traveling with a compass. If you're consulting a map, place the map on a flat surface and place the compass on the map to get the most accurate reading.

Find the Direction You're Facing

For a quick basic exercise in orienting yourself, you want to find out which direction you're facing or traveling. Look at the magnetic needle; it should swing off to one side or another unless you're facing North.

Turn the degree dial until the orienting arrow lines up with the magnetic arrow; both pointing North. Now find the general direction you're facing by looking at the direction of the travel arrow. If the travel arrow is now between the N and the E, you're facing Northeast.

Find where the travel arrow intersects with the degree dial. To take a more accurate reading, look closely at the degree markers on the compass. If it intersects at the 23, you are facing 23 degrees Northeast.

True North and Magnetic North

While this may seem confusing, there are two kinds of "North." However, you can learn this distinction quickly, and it is an important piece of information in order to use a compass properly.

True North or Map North refers to the point at which all longitudinal lines meet on a map or the North Pole. All maps have the same layout with True North at the top of the map. However, due to slight variations in the magnetic field, your compass doesn't point to True North; instead, it'll point to Magnetic North.

Magnetic North is the tilt of the magnetic field or about eleven degrees from the tilt of the Earth's axis. This means the difference between True and Magnetic North is as much as twenty degrees in some places. Depending on where on Earth you are, you'll have to account for the magnetic shift to get an accurate reading.

While this difference may not seem that big, traveling just one degree off for a distance of a mile will make you about one hundred feet off track. Now consider you hike for ten or twenty miles and how many feet off track you'll become. This is why it is important to compensate by taking the declination into account.

Understanding and Adjusting for Declination

A compass arrow points to the magnetic north while a map arrow indicates true north. The difference in these two is known as declination and is measured in degrees. Every year the declination gradually changes. For example, in the United States, the amount of declination can vary greatly from a position on the East coast to a position on the West coast.

Correcting for Declination

To make using your compass easier, you can correct for declination by either adding or subtracting the declination amount from your bearing in degrees. This depends on whether you are taking a bearing from a map or from a compass and whether you're in an area with East or West declination.

In the US, the line of zero declination runs up through Alabama, Illinois, and Wisconsin at a slight diagonal. East of the line, declination orients towards the West. West of the line, the opposite is true. Find the declination in your area in which you plan to travel so you can adequately compensate for it.

Actually Using a Compass

Find the Direction You're Headed

No matter how well aware you are of your wilderness location, it is good to periodically check your bearings to make sure you're still headed in the right direction. To do this, move the compass until the direction of travel arrow is pointing in the direction you've been going and will continue heading. Unless you're heading north, the magnetic needle will spin to one side.

Twist the degree dial until the orienting arrow lines up with the north end of the magnetic needle. Once these are aligned, you will know where your direction of travel arrow is pointing.

Take off the local magnetic variation by twisting the degree dial that correct degrees either to the left or the right, depending on declination. Note where the direction of travel arrow lines up with the degree dial.

Continue moving in this direction. To do so, simply hold the compass in the proper stance and turn your body until the north end of the magnetic needle aligns with the orienting needle again and follow the direction of travel arrow.

Check your compass as often as needed, but make sure you don't accidentally move the degree dial from the current position.

Focus on a point in the distance. To accurately follow the direction of travel arrow; look at the arrow and then focus on a distant object to use as a guide. Don't focus on anything too distance such as a mountain since huge objects aren't precise enough to accurately navigate.

Once you reach each guide point, use the compass to find another. If you have limited visibility and you can't find distant objects, you can ask your fellow Bushcraft practicer for help.

Transpose Direction of Travel to a Map

First, place your map on a horizontal surface and then place the compass on the map, so the orienting arrow points to true north on the map. If you know your position on the map, move your compass around so that the edge passes through your current position, but the orienting arrow should continue to point north.

Draw a line along the edge of the compass and through your current position. If you keep this bearing, your path from your current position will be along the line that you just placed on the map.

Take a Bearing from a Map

In order to find which direction you need to travel to get somewhere, place the map on a horizontal surface and place your compass on the map. Use the edge of your compass as a ruler and place it, so that it creates a line between your current position and where you want to go.

Rotate the degree dial until the orienting arrow points to true north on the map. This also aligns the compass's orienting lines with the north-south lines on the map. After the degree dial is in place, you can put away the map.

In this case, you make the necessary correction for declination by adding the appropriate number of degrees if you're in the West declination and subtracting in areas that have an East declination. This is the opposite of what you do when taking a bearing from the compass.

Hold the compass horizontally in front of you with the direction of travel arrow pointing away from you. This arrow will guide you to your destination. Turn your body until the north end of the magnetic needle is aligned with the orienting needle and you'll be oriented toward your destination on the map.

FINDING YOUR BEARINGS WHEN LOST

First, you want to find three prominent landmarks that you can see and also find on the map. Finding where you are when you don't know your location on a map is one of the more difficult tasks with a compass, but it is also one of the most important.

You can re-orient yourself by finding three distinct landmarks spread around you in as wide a field of view as possible.

Next, aim the direction of travel arrow at the first landmark you see. Unless the landmark is to your north, the magnetic needle should spin to one side. Twist the degree dial until the orienting arrow lines up with the north end of the magnetic needle.

After they've aligned, you will know where the direction of travel arrow is pointing. Now correct for declination based on your area.

Now you need to transpose the direction of the landmark onto the map. Place your map on a horizontal surface and place the compass on the map, so the orienting arrow points to true north on the map. Now slide your compass around so that the edge passes through the landmark on the map with the orienting arrow continuing to point north.

Lastly, you can triangulate your position. Draw a line on the edge of the compass and through your approximate position. This will be the first of three lines you will draw from all three landmarks. You can then find your position by forming a triangle with the other two landmarks.

Repeat the above process with the other two landmarks. Once you are done your map should have three lines that form a triangle. Your position is within the triangle, the size of which will be based on how accurate your bearings are.

What If You Don't Have a Compass?

The first thing you need to remember is that the sun rises in the East and sets in the West. Depending on the time of day, you might have to wait a little bit to determine this information.

As long as you know this, an analog watch can be used to get a heading if you don't have a compass. Position your watch in front of you. Turn it so the hour hand or short hand points towards the sun. Holding it this way, noon will be south. If the time is before six am or pm then divide the angle in between the 12 o'clock and the shorthand in half and make that way south.

If you don't have a compass, you can also look to nature for help. The north side of trees typically has fewer branches and more moss or lichen. It is also a known fact that ants prefer to make nests on the south side of trees.

Also in the spring time, snow melts faster on southern facing slopes or southern rock faces. Undergrowth is often thicker on slopes that face south. However, these indicators aren't 100% reliable, so you need to check with as many ways as possible to make sure you are heading in the right direction.

If you are in the Northern hemisphere, the Polaris star is approximate North at all time. It is easiest to find the Big Dipper first. Find the two stars near the end of the constellation and then draw an imaginary line up and about five times the length of the distance between the two stars. You should be able to see Polaris and know that direction is North.

On the other hand, if you are in the Southern hemisphere you can use the Crux or Southern Cross to determine south. This is a group of four stars close together.

While it is named the Crux or Cross; it actually looks more like a kite. It can often be located at different points in the sky. Depending on the season is can be low in the southeast or high in the southwest.

Other than a compass, the most accurate method requires a clear sky and a lot of time. You can create a makeshift sundial with a straight pole about a yard in length, two smaller sticks and a piece of string. At noon the shadow of the stick is going to be to the south.

5 Common Compass Mistakes to Avoid

When using your compass near any metal object interference can occur. This includes weapons, belt buckets, zipper pulls and even pens or batteries in your watch or phone.

1. Forgetting to orient the compass to the map before determining the direction of travel.
2. Not holding the compass properly.
3. Watching the compass while you walk rather than taking a visual bearing on a landmark or object and walking to that point.
4. Not checking your bearing often enough.
5. Not accounting for declination or using an outdated or incorrect declination for your specific location.

While you may not use your compass regularly while practicing bushcraft skills, it is still a vital piece of equipment to have with you. You certainly don't want to be caught without a compass when you need it.

Take the time to learn and practice both map and compass skills to improve your navigation as a part of your bushcraft skills. Now you are ready to head out on your first trip and get started practicing your bushcraft skills. Let's see what you need to do on your first trip into the wilderness.

YOUR FIRST TRIP

The first step to take when planning your first bushcraft trip is to find the appropriate location. Many public parks and campgrounds have rules and limitations on what you can do.

For example, most campgrounds won't allow you to gather firewood. You often won't be able to go on private land unless you know the owner and have permission to use their land. So where are you supposed to go to practice your bushcraft skills? There are a few options you can do.

The first option is to simply fly under the radar at a public park. However, I don't recommend this as you are doing something illegal and could be faced with a fine. If you are going to do this then there are a few things you need to do:

1. Don't light a fire.
2. Have a signaling system if you happen to get lost.
3. Tell someone where you are going and when you expect to return.
4. Be discrete and clean up after yourself.

Another option is to find some private land that is suitable for bushcraft and then asking the owner if you can use it. It is best to be open and ask outright. When you ask, be sure to highlight the following five things:

1. You won't hunt.
2. You won't trap unless you ask for permission first.
3. You won't make a fire unless they're okay with it and if you do you'll take all necessary precautions.
4. You'll tell them before you go on their property.
5. You'll give them contact information in case they need to reach you.

Lastly, the third option is to find lands that are owned and maintained by one of two organizations and negotiate with them for use.

1. The Boy Scouts of America have a number of camps which you may be allowed to use in the off-season.
2. The civil air patrol owns and operates lands that they may allow you to use.

FIRST TRIP PREPARATION

Once you know where you are going to practice your bushcraft skills, you want to make sure you have all your appropriate gear packed. Eventually, you will only be bringing a knife with you, but on your first trips, you should bring some backup gear in case you need it.

Make sure you Google the declination for the area you are going to be in so you can properly practice your compass.

You should also make sure you have a good topographical map of the area you are going to be practicing in so you can get map reading skill practice. Practice identifying landmarks such as rivers, ridges, valleys and high points as well.

FIRST TRIP STEP BY STEP

Once you arrive at your destination, you can start practicing your bushcraft skills. As we've learned, you want to start out with the four survival basics. Let's see what you should do when on your first trip outdoors. The order in which you do these things might change depending on where you go and what season you are in since you'll need to prioritize based on your needs.

In general, your first step is going to be building a shelter. As we've discussed, either look for a natural shelter that can sustain you for the short-term while you're outside practicing bushcraft or take the time to actually build a long-term shelter. The main thing is to be both protected from the elements and get practice in building survival shelters.

Next, provided it is allowed where you are, you should practice starting a fire. Practice various methods or if you have already done this in your backyard; just start a fire to improve your abilities.

Then you should practice looking for water sources and setting up a water filtration and purification system. Always bring some purification tablets with you or extra bottle water if you're only going to be out for a night. This way if you can't successfully find and purify your own water, you will still be able to survive.

Lastly, you'll want to focus on your food. If it is allowed where you are, try hunting and trapping food. If you aren't in a place to do this, at least focus on tracking and locating food sources and foraging. If you are able to hunt and catch food then also work on practicing your butchering and food preparation.

Try to practice both new skills and old skills when you head outside. Even if you've been practicing at home in your backyard, nothing compares to using it in a real situation. Whether you are on your first trip or at home, look at some activities you can do to practice your bushcraft skills.

BEGINNER BUSHCRAFT ACTIVITIES YOU CAN DO

As you can tell from reading this book so far, there are quite a bit of things to learn when it comes to bushcraft. Whether you are on your first trip or just getting out in your backyard to try some simple beginner bushcraft skills, consider trying the following.

FOOD FORAGING

Take a hike in the woods and look for edible plants. As you hike, make sure you try to identify as many local plants as possible. If you find one that is on the edible list, give it a try. This way you'll know what to identify when you really need it.

Trapping and Hunting

Also while hiking try to find and identify at least one set of animal tracks. When you get home read up on the animals and see if you were correct in your identification. Also, read up on local animals to get yourself prepared for the next trip.

Water Harvesting and Purification

Try your hand at locating water signs while on your hike. Then when you head out into the wilderness, you can practice harvesting and purification. You can also try some purification methods at home in your backyard. This is an important skill, and you want to practice it often until you are comfortable.

BUILDING SHELTERS

This is an extremely helpful bushcraft skill but is difficult to practice in your backyard. However, this is an easy skill to practice when you head out into the wilderness.

You should take the time to determine what type of shelter is best for your area, but still, learn how to build all types of shelters, so you are prepared for any type of situation or disaster.

FIRE BUILDING

Due to local laws, you may have to practice this one in your backyard only. However, you can still practice locating flint, tinder and firewood while out on a hike or during your first trip. You can never spend too much time practicing how to start a fire.

Finding a Survival Location

Everything we have discussed so far is based on an ideal world where you have time to practice and prepare. However, we all know a survival situation can happen tomorrow. Therefore, you also want to consider some real life scenarios in case something comes up, and you suddenly find yourself in a survival situation before you've completely practiced your bushcraft skills.

The most important thing is to know where to go in a survival situation. If you find yourself needing to know where to go and look for the following:

Make sure to choose a location that is off the beaten track and ideally only accessible by a single road. This seclusion can offer you an added degree of protection. For example, you can cut down a tree across the road to effectively prevent people from coming to you.

When you start setting up a long-term shelter, don't get too ostentatious. You don't want to draw attention from others and become a target for vandalism or stealing. Stay simple and solid in your long-term shelter plans.

You should ideally choose a location near a natural water source whether it is spring, well or stream.

Make sure you set up at least one fireplace or stove for both heat and cooking.

You still want to be at least ten to twenty miles from a village or small town, at least someplace you can travel to on foot if needed. This way you have someplace to go should you need emergency supplies, get updated on the news or otherwise get some contact with the outside world and other people.

If it is going to turn into a true long-term survival situation you want to make sure you have arable land to grow vegetables and other crops.

Along with growing your own food, you also want to be close to other food sources such as hunting and fishing.

Ideally, the location should be able to provide enough food to keep you stocked for three months at the least, but ideally up to a year.

This is a lot of information, and you're going to need to practice and take your time learning. However, the last area to look at is how to save money when getting prepared to start learning bushcraft skills.

MONEY SAVING TIPS

Obviously, once you have mastered bushcraft skills, your only purchase is going to be a knife and maybe another bladed tool. However, when you are first starting out you are going to pack some gear with you, and you're going to want to save as much money as possible.

Especially if you are only going to use this gear a few times before heading out into the wilderness without it. So let's see how you can get all the survival gear you need for cheap.

CLOTHING

Clothing is important and if you are already an outdoors person then you likely already have what you need. However, if buying outdoor clothes are still new to you, then there are a few things to consider.

First, you never want to buy clothes based on claims that are too good to be true. Most materials and designs will work well for outdoor use. The only material that is generally avoided outdoors is cotton. Cotton dries slowly and isn't very warm when wet.

When choosing clothing you want to approach it in layers. You should always layer several thin garments rather than one thick garment so you can easily thermally regulate. Each layer has a specific function that it needs to make.

The base layer directly against your skin should be a material that pulls sweat away from your body. For the cheapest option, you can buy a simple synthetic or polyester t-shirt. Often you can get these at a standard store like Target for about $5.00.

Next is the mid-layer. When it comes to pants, your best value is a pair of cargo pants that is at least 50% synthetic. Most department stores carry them, or you can find a decent new pair at a military surplus store for about $20.00.

For the upper body, you would want something like a thin fleece. You can also go with a long sleeve synthetic shirt. Most stores like Target and Wal-Mart have good fleece options for about $20.00 to $30.00.

Lastly, the outer layer needs to stop the wind and also protect you from possible rain. For this, you can go with a simple nylon jacket. You don't need a jacket with insulation since this is provided by the mid layer. Most stores will have a simple nylon jacket for around $30.00.

For socks you want a good wool pair to make hiking more comfortable. The socks don't have to be complete wool, and they don't need to be any specific type of wool. You can often get a three pack for about $20.00, or you can get them for cheaper at most Army surplus stores.

BACKPACK

A backpack is something you are going to need in order to carry any gear you bring. It is best to buy all your other gear first, so you know what you need to pack and it will make it easier to choose a good backpack. A good quality backpack is going to be one of your more costly purchases. You can expect a hiking backpack to be about $100.00 to $150.00. However, if you really want to reduce your cost, you can consider going to a military surplus store and get an ALICE pack which can be as cheap as $30.00 or less.

SHELTER

You probably don't need to spend a lot of money here, if any. With practice, you can make your own shelter in the wilderness, and you won't need to pack a tent with you.

However, if you want to have a leg up on building a shelter in the wilderness until you get comfortable building one from all natural materials; consider at least buying a tarp.

There are some hardware stores where you can get tarps for $10.00 or less. A tarp can provide a quick and simple shelter that is far cheaper than buying a tent.

Another item that you may choose to carry is a sleeping bag. Even after you get experience with bushcraft skills, you may still choose to carry a sleeping bag for added warmth. While there is no low-cost option when it comes to sleeping bags, you can lower your cost options.

First, you want to avoid rectangular sleeping bags. These are cheap and often comfortable, but they don't give you the best insulation.

Rather you should spend your money on a mummy style bag since this will fit close to your body and preserve the most heat. When it comes to buying sleeping bags, you will notice two main types.

There is synthetic and down insulation sleeping bags. Each comes with advantages and disadvantages. However, we are primarily concerned with cost. Synthetic bags often have the advantage in this category, but they are going to weigh more. Still, you can expect to pay out $100.00 to $150.00 for a decent sleeping bag.

ROPE

This can help you perform several general tasks until you get into the advanced bushcraft skills of making your own rope. Your best option is paracord which is a strong and thin rope. Typically you can get about 100 feet for around $4.00 at an outdoor supply store.

COOKING POT

Until you get good at cooking without utensils, you might want to at least bring a cooking pot with you. Ideally, you should choose a pot that provides a good balance between cost, weight, and performance. You can often find a decent option for about $10.00.

WATER FILTER

There are a number of water filters to choose from these days. It is always a good idea to have a few on hand until you are comfortable with water purifying methods in the wilderness. You can often find water filters for about $25.00.

However, once you get adept at collecting water you would only need to boil it before drinking it, so you may not need these too often.

KNIFE

This is the area where you don't want to skimp on the money. A good quality knife is the key for bushcraft. Not only are you going to use this for pretty much every activity in the wild, but you are also going to be using it for several years to come. So I would spend my money on getting a wonderful knife that is going to truly last awhile. These can easily cost you $100.00 to $200.00.

Depending on what you choose and where you shop you can easily get your basic supplies for under $500.00. However, there may be additional supplies not listed here that you would prefer to carry or may be needed in your specific area.

Be sure to make your selection carefully and keep in mind that the more experience you gather over time by practicing, the fewer tools you will need to survive in the wilderness.

Now that you know what you need to learn, and you have everything you need, get out there and start practicing your bushcraft skills. I'm still mastering bushcraft, but you can probably learn the basics within a year and be well prepared for an outing in the wilderness.

Made in the USA
Monee, IL
04 November 2026

67388645R10046